OPEN AND
INDUSTRIALISED BUILDING

CIB Publication 222
Report of Working Commission W24

OPEN AND SEBE
INDUSTRIALISED BUILDING

Edited by

Asko Sarja

Professor, D.Tech.
Technical Research Centre of Finland
Espoo, Finland

International Council for
Building Research Studies
and Documentation

E & FN Spon

An imprint of Routledge

London and New York

First published 1998
by E & FN Spon, an imprint of Routledge
11 New Fetter Lane, London EC4P 4EE

Simultaneously published in the USA and Canada
by Routledge
29 West 35th Street, New York, NY 10001

Printed and bound in Great Britain by
St Edmundsbury Press, Bury St Edmunds, Suffolk

British Library Cataloguing in Publication Data
A catalogue record for this book is available
from the British Library

ISBN 0 419 23840 9

Contents

Preface

The objectives of this book are to analyse the current status and state of the art of the industrialisation in building in different regions of the world, and to present general conclusions and trends along with some personal visions of the future developments.

The report is the result of a project carried out by CIB Working Commission 24 "Open Industrialisation in Building" during 1994-1997. The state of the art review is based on the results of a questionnaire sent to experts world-wide, on reports of CIB W 24 members during Commission meetings, on available literature, and on the personal experience, knowledge and conclusions of the author. The literature is selected from original reports from different countries. The world is divided into eight regions: European Union, Central and Western Europe, Russia and GIS countries, Northern Asia, South and Southeast Asia, North America, South and Central America, Middle East and Africa. Although quite large, each region has fairly similar environmental and socio-economic conditions, with resulting similarities in the development of building technology. Some countries from each Region have been chosen as representative samples for description and analysis. The amount of information is so vast that strict selection was necessary to ensure compactness and uniformity of the report.

This work focuses on analysis of, and conclusions concerning, the general state of development and current development trends. For this reason the report has been kept as concise as possible to give the reader the core aspects of the subject. In the conclusions I attempt to explain earlier development paths, current states and trends of the development per region, as well as the relationship between the current state and development and the history, traditions, technical level and socio-economic characteristics of each society. These factors will largely dictate predictions of future developments. Although we have strived for objectivity many of the conclusions are naturally subjective, and the critical reader may differ in his or her judgment of the result.

The second part of the report deals with future visions of building. Several experts and researchers share their own visions of future building technology and the role that open industrialisation may be expected to play in it.

I hope that this report will contribute to the discussion on the current state and future development of building technology in widely differing conditions around the world, possibly serving as a launching pad for new ideas in research and development as well as in practice.

Professor D.Tech. Asko Sarja
Espoo, Finland, November 1997

Acknowledgements

The editor wishes to thank all authors referred to in the book, and members of CIB W24 "Open Industrialisation in Building" for their contributions, comments and personal visions. Many thanks also to Dr. Colin Gray, who make a mammoth effort in unifying the English language of the literature reference text, to Mrs Adelaide Lönnberg for her skilled checking of the language in the editor's texts, and to Mrs Soili Miekkola for the systematisation references.

Part 1

State of the art and trends in open industrialisation

Asko Sarja

Professor, D.Tech. (Finland)

1

Background

Building production world-wide is a vast business area, which is important for the entire welfare of the people, for the economy of organisations and companies and for general sustainable development of societies. The societies in most developed countries have already passed from the industrial era into the post-industrialised phase, while the NIC countries are at a fast developing phase of industrialisation and the developing countries are at the very start of it.

Building technology is still mostly a local production, which applies industrial principles quite inefficiently. This is one reason why housing apartments and buildings for business and industry are generally expensive and production is sensitive to recessions. Building production has been gradually developing towards industrialised production, already over several decades. Commonly used prefabricated products are small structural components like doors, windows and kitchens as well as heating and building services and electrical products. The bearing structural system, partitions and facades vary widely in the share of prefabrication in different countries.

The construction process and project management methods have generally undergone quite little industrialisation compared with industrial production in other fields, such as in the automotive industry, shipbuilding and mechanical industries. It can be stated that building technology as a whole is a semi-industrialised production area while the other fields of industrial production have already moved into the post-industrialised information era.

We are recognising that the slow increase in productivity of the building sector has already led to increasing costs, and that therefore the building market in developed countries lies in a long term stagnation phase. An important tool in re-activating the building market is the reduction of price levels. In the NIC countries the need for building is high and the prerequisite to effective industrialisation is the increase of production capacity. In the developing countries a gradual increase of productivity in manual working through semi-manual equipment and through industrial production of some bulk products is the main demand.

During the mass production era of the 1960s and 1970s the first attempts toward wider industrialisation of building were made in industrialised capitalistic and socialistic countries. Production requirements dictated town planning and architecture at the expense of social and aesthetic considerations. The technical quality was often insufficient especially in socialistic countries, with resulting general bias against industrialised building.

In order to meet economic requirements together with other requirements like aesthetics, functionality, ecology, durability, health and comfort of living, we have to rethink the entire content and solutions of industrialised building. Increasing international exchange of technology and building products activates industrial development through increasing competition and evolution with the help of a continuous selection of the most competitive technologies and products. This requires international openness of building, product and information systematics, because only in this way will we activate and accelerate the internationalisation trend. Buildings play a major role in defining the local culture of our built environment. Therefore even international technologies and products must be highly adaptable to both local traditions and the natural environment.

2

Glossary and definitions

2.1 GENERAL DEFINITIONS

Today the industrialisation of building means the application of modern systematised methods of design, production planning and control as well as mechanised and automated manufacturing processes.

The required openness refers to the capability to assemble products from alternative suppliers into the building and to exchange information between partners of the building process and inside the consortia and business networks. The application and exchange of products, services and information nationally and internationally as well as the adaptation of the products and services into varying local needs and cultures is essential. For this purpose an effective international co-operation is needed in order to develop proposals for definitions, rules and models for this kind of regional and local open building, utilising global technologies and methods into local applications.

The overall system is aimed at forming an entire building from interacting items. **The system** thus can be defined as an organised whole consisting of its parts, in which the relations between the parts are defined by rules. The system can be a **product system, an organisational system or an information system**. In the **open industrialised building** product system, the organisational system and information system are bound together.

The central scope of open industrialisation includes the following areas:

- Demand Side, dealing with user requirements and with the introduction of the requirements into designs.
- Supply Side, dealing with the production requirements and with the linking of demand and supply.
- Building Process, including organisation and communication in building projects.

Open system building is a general framework for the building industry, including modular systematics of products, organisation and information, dimensional co-ordination, tolerance system, performance based product specifications, product data models etc., so that the suppliers serve products and service modules that will fit together.

Openness is a concept with many aspects, like:

* OPEN for individual designs
* OPEN for competition between suppliers
* OPEN for alternative assemblies
* OPEN for future changes
* OPEN for information exchange
* OPEN for integration of modules and subsystems.

2.2 SPECIFIC DEFINITIONS

2.2.1 System

In this chapter several more or less similar existing definitions are presented in order to explain the term "System". It is important to realise that most of these definitions are generally used in other branches of industry and play an important role in the development of modern industrial principles, methods and products, and materials and information processes. This kind of systematics was used in limited applications already in ancient times for masonry and stone structures, but the theoretical development started in the furniture industry and was followed by the so-called "Bauhaus" development in Germany in the 1920s. Later the centre of this development shifted to the mechanical industry and, in recent decades, to electrical and computer theories and businesses [46]. Even music composition and languages can be treated as modular systems having the same theoretical principles and in some cases terms as in technical and information systems. Some selected definitions are as follows:

A system is defined as a set of parts with holistic potential; that is, a set of parts organised to act as a whole [12].

A system is an organised whole consisting of its parts, in which the relations between the parts are defined by rules. The parts can be concrete (e.g. the components of a building system) or abstract (e.g. the components of an information system) [46].

System: An organised entity consisting of components which have defined relationships [46].

A system is usually defined as a group of interrelated components that act together to accomplish a specific objective [52].

System: a set of objects with relationships between them and between their attributes (comment: the boundaries of a system will change depending on the relationships being considered; thus, considering a building system from a technical point of view will lead to a study of components, joints, tolerances etc., and considering the same system from an organisational point of view will lead to a study of components, joints, tolerances etc., whereas considering the same system from an organisational point of view will involve the consideration of procurement, management, marketing etc.) [19] (Davidson).

A system is a set of entities, which may be real or conceptual (material or immaterial), and these interact with one another to form a whole: an open system has a specified environment within which to react whereas a closed system is one that ignores its environment. In the open system energy can pass in and out but the system maintains its homeostatic state - a principle enunciated by Cannon in 1939 [45].

2.2.2 Building system, Open building system, Closed building system

In this chapter, the following selected definitions are presented:

Building system: This includes design rules and a product system whose parts have compatible interfaces thus permitting the use of several alternative components and assemblies. The compatibility of the components and assemblies is assured by means of a dimensional and tolerance system as well as of connections and joints [46].

Set of building components: A set of components designed in such a way that they can be combined with each other in different ways and that the links with other components are studied in advance. The components of such a system can be described in a "System Catalogue", the rules for combination in a "Design Guide" (grids, rules for positioning, performances,…) [15].

Open system: A system that does exchange objects or information with its environment (comment: an "open system of building" can be envisaged in technical terms, provided certain conventions are respected; in organisational terms, there is evidence that "open system building" presents a number of disadvantages that can conveniently be overcome with some organisational "closing" of the systems) [19] (Davidson).

Open system: The integral open system is utopian, as is a universal assembly system. The bases of the "open system" which are possible under such conditions are as described below:

There are catalogues of components which comply with the same dimensional discipline. These components have assembly compatibility which is not always necessarily identified. The hypothesis is that a set of catalogues of locally available catalogues contains sufficient compatibility and diversity to satisfy a reasonable demand for architectural and technical variety [19] (Lugez).

Open system buildings: These are buildings that have been put up with components coming from independent producers, each with his own catalogue, the catalogues altogether forming the general catalogue of the open system [10].

'Systems to order': Systems using components manufactured to order, which may be called 'systems to order' (as an example of this, almost all present systems of concrete prefabrication) [10].

"Open system building" or "Open Industrialisation of the building sector": An approach for the construction sector providing a framework so that different "Open building systems" can be elaborated. The core of this framework are standards or rules on dimensional co-ordination and compatibility of performances [15].

Openness of the building system: The system is open for

- free design for varying requirements
- free competition between contractors and suppliers
- future changes in the use and
- reuse and recycling [46].

Open Building is a set of principles for constructing and renovating residential and other kinds of buildings. The principles help solve problems when decisions are organised on several levels, among a number of parties who prefer to act independently while expecting a coherent architecture to result [29].

A system is a set of entities which may be real or conceptual (material or immaterial) and these interact with one another to form a whole: an open system has a specified environment within which to react whereas a closed system is one that ignores its environment [45].

A closed system is one in which there is no import or export of energy – the common example given is a number of reactants brought together in a closed vessel. However, truly closed systems are hard to find and it is most useful to think of the terms as relatively descriptive, and in this way they have a purpose. For a system to be identifiable at all it has to have a degree of closure and it is again what constitutes that closure that is of interest [45].

Closed system: A system that exchanges nothing with its environment (comment: in reality, a "closed system of building" cannot exist in technical terms, since at least the cement, reinforcing steel, plumbing fixtures and paint will be brought in from outside the bounds of the system; in organisational terms, however, a "closed system of building" denotes one which is under a single, unified controlling organisation with which all clients must deal for virtually all aspects of a project- and it is this aspect which is seen (a) to limit the market for the parts to the market of the whole, and (b) to constrain the freedom of decision-making of the ex-system participants, notably the architects) [19] (Davidson).

The term "building system" or "system building" as interchangeably used means "a generating system for building or a set of building components which may be assembled in different ways to create a variety of building configurations" [27] (Commandante).

A building system is defined as a set of components for a particular type of building, together with their production and erection procedures [52].

Meccano set: Systems using components taken from a firm's catalogue, the components having been designed for assembly among themselves; this is what is known as the meccano set by analogy with the construction toys that also include a catalogue of components [10].

Editor's comments:

The various definitions above are mostly quite similar, the differences reflecting different phases of development of systems. The earlier way of thinking concerns only the compatibility of components, which often are understood as standardised catalogue components. The latest definitions point out more general systematics, including product system, organisational system and information system. Instead of strict standardisation of products, the compatibility rules as well as compatible geometry at interfaces and performance properties are important.

2.2.3 Hierarchical and modular building system

The following definitions are selected:

Hierarchical building system: A building system consisting of hierarchical subsystems where each part of a lower level belongs to only one part of the upper level. The degree of a system is the number of its hierarchical levels, i.e. combination levels [46].

In a hierarchical system the parts can be located at different levels in the organised whole. The parts of an upper hierarchical level are composed of the parts of the hierarchical levels below it [46].

An open modular system consists of modular parts at different hierarchical levels, for which it is possible to produce different interchangeable products and designs that can be joined together according to connection rules to form a functional whole [46].

Modulation involves division of the whole from the perspective being dealt with at any particular time (e.g. functions, spaces, technical design, production planning, quality) into sub-entities, which to a significant extent are compatible and independent. The modulation can take place at different hierarchical levels [46].

The parts of a modulated system can form at different hierarchical levels sub-entities, which to a significant extent are independent. The relations between the sub-entities are defined by the system's rules [46].

2.2.4 Industrialised system building

A special group of systems, often referred to as "industrialisation onsite" or simply as "industrialised building", combines intensive utilisation of various precast elements with highly rationalised framing construction methods [52].

Industrialisation process is defined as an investment in equipment, facilities, and technology with the purpose of increasing output, saving manual labour, and improving quality [52].

Industrialisation is the use of technologies to replace the skill of the craftsman by using a machine. It fits in naturally in a workshop or factory, but also on-site when a machine may be used to replace a craftsman. Industrialising can involve either:

- Series production: Productivity results from the repetition of identical operations for the manufacture of products which are themselves identical, or
- Recourse to automation: Products are diversified within a certain range, without any loss of productivity. For this purpose, the production tool is adapted to the order it receives for each of the products to be manufactured [19] (Lugez).

The term "industrialised" refers to parts fabricated by machines and organised production in labour. Within this definition of terms, an industrialised building system, therefore, is defined as a set of building parts which are mass-produced [27] (Commandante).

Industrialised building is the building technology where modern systematised methods of design, production planning and control as well as mechanised and automated manufacturing are applied [46].

Today, open industrialised building means the exchange and application of products, services and information between the players in the building process, nationally and internationally, and between countries and regions with major differences in the structure of the building industry [3].

2.2.5 Prefabrication

Prefabrication of product systems: Presently applied systems of prefabricated building components and building parts are in some quarters considered as open systems. Prefabricated components for frames and envelopes from various manufacturers can, certainly, be made compatible – after costly adjustments either in the factories or on the building site. This can be seen as one of the reasons why during the 1990s the share of "semi-prefabricated" composites (prefab + in situ) in the frames of apartment buildings grew almost to the double the share of fully prefabricated components. Prefabrication is a buildings production where the components or assemblies are manufactured fully (full prefabrication) or partly (composite construction) in factories and assembled together on site [46].

2.2.6 Subsystems, modules, components, materials

Subsystem:

A system which forms a part of the building system. The subsystems of the building are:

1. The architectural planning and design systems: The functional system of the use and the spatial system serving the functions of the use
2. The technical systems: Structural system and installation systems (Heating H, Ventilation V, Cooling C, Electricity E, Water W,

Sewage S, Information I, and Waste Management M). Each subsystem consists of the hierarchical levels building, sub-building, module, component and material [46].

Modulus: To favour series production, the dimensions of spaces are chosen from multiples of an increasing pitch referred to as the modulus (M = 100 mm) [19] Lugez.

Semi-products are materials that have been worked up and given some shape, but that are not intended for a particular use in building. This heading may cover sections which have a quite definite cross section but no specified length; it may include slabs, such as an asbestos-cement slab; it may be concerned with volume, like a brick or block. We advise calling these materials "semi-products" for, in fact, they do not have a unique use in building but may in contrast be put to a variety of uses [10].

The component as such is a product that has been designed for a particular use in building, or for a limited number of uses [10].

A cell is referred as a design element of room-size dimensions, which can be used as a building block for different layouts [52].

Linear systems use as their main structural elements columns, beams, frames, or trusses made of plain or prestressed concrete. Their important feature is the capacity to transfer heavy loads over large spans [52].

2.2.7 Dimensional system and dimensional modulation

The reference system consists of planes, the intersection of those planes (lines) and the intersection of those lines (points). The reference system is used to define the position and size of components and buildings [16].

A rectangular reference system consists of parallel or orthogonal planes [16].

A grid is a set of lines defined by the intersection of a spatial reference system and a selected plane [16].

2.2.8 Infrastructure and processes of open industrialisation

Perspectives, e.g. design, manufacturing, production, use, quality and service life, are the bases of the rules and modulation of the system's different hierarchical levels and lead to different definitions [46].

Horizontally integrated production: Mainly through the open building system and open design horizontally integrated companies marketing alternative, mutually compatible deliveries (materials, components, modules, assemblies and services) for the contractors or clients thus competing with one another [46].

Hierarchy of deliveries in building projects:

- Materials delivery, including defined materials
- Component delivery, including defined components
- Module (assembly) delivery, including a defined, mainly independent set of components and their detailed design
- System delivery, including an entire technical system
- Total system delivery, including the technical system, its detailed design and assembly
- Building delivery, including the construction of the entire building
- Total building delivery, including the architectural design and construction of the entire building [46].

2. 2. 9 Open design

Open computer aided design: design organisation and rules, where the partners can transfer the data in real time between each other. The openness includes the rules and system rules of general data bases, of data structures in design and of standard data interfaces between the partners. A general product data modelling manufacturing, construction, marketing, use and management of the building, repair and recycling [46].

3

State of the art in open industrialisation

3. 1 GENERAL STATUS AND TRENDS OF INDUSTRIALISATION

The technical, architectural goals and realisation of industrialised building changed most during the 1980s and 1990s. The current technology is flexible for individual architectural designs also allowing easy alterations during use, future changes, and modernisation. For this aim, systematic modular design and products are applied, including dimensional modularity, tolerance system, compatible joints and use of modular products. As technical tools, highly mechanised and automated methods are increasingly applied to the manufacture of components in factories, as well as in assembly and some phases of on site manufacturing. Computerisation is increasingly applied in the entire building process: in analysis, design, manufacturing, production and operation.

The current phase of development of open industrialisation world-wide can be investigated only by analysing the different sub-areas of open industrialisation separately. Therefore, CIB Working Commission W24 investigated the status of open industrialisation as a whole in several countries from different parts of the world.

The status reports for some countries are summarised in Table 1. Advanced open system building is dominant in some countries like Finland, the Netherlands and Denmark. In other countries like Japan, France, Germany and the USA, closed systems based on company specific applications predominate. In countries where open systems dominate, the market share of modern industrialised production is high, whereas in countries with closed systems the rate of industrialisation is considerably lower. This indicates that open systems strongly undergird industrialisation.

Table 1. Status of open industrialisation in some countries.

Country Feature	Finland	India	Israel	Japan	Sweden	USA
1. General	Dropped volume New buildings mainly in the largest towns Small sites Pre-fabrication dominant Value of construction 8% of GNP	Vast shortage of houses Most population in the countryside (73%). Wide variety of technology and materials	Rapid growth of population, building and economy. Multi-family houses dominant. Value of construction 12% of GNP	Traditional buildings wooden structures. Large production. Great variety in the size of buildings (from small to sky scrapers.	Dropped volume. Smaller projects. Value of con-struction 12% of GNP	Multifamily 10.5%, 3.5% HUD code. 86% single family.
2. Materials	One family houses: Wood Detached houses: Wood and concrete Multi-storey apartment: Concrete Offices: Concrete, some steel	Rural: Local natural materials. bamboo, earth, timber, stone, bricks Urban: Bricks, concrete, stone	Concrete dominant. Small share of lightweight steel or timber.	One family houses: timber, concrete. Multi-storey apartments: concrete Offices: concrete or steel	Single and detached houses: Wood Apartment: Concrete Offices: Concrete, some steel.	Single houses: Wood Multifamily: Concrete, steel Offices: Steel, concrete

3.2 STATE OF THE ART IN SOME REGIONS AND COUNTRIES

3. 2. 1 European Union, Central and Western Europe

In European Union countries the industrialisation of building production started early. The pioneering research work was done already in the 1920s and 1930s in Bauhaus, Germany, where entire open building systems were proposed and experimented with. Several architects from Bauhaus, like Gropius, later worked in the USA developing highly industrialised building concepts for high rise buildings.

Generally, the European industrialising process in building has progressed quite slowly through increasing use of industrially produced building products, components and manufacturing equipment. In the 1960s the vast mass production of buildings needed effective standardisation and industrialisation of structural components and entire building systems. To meet that need many different systems were developed and used.

In Western Central and Southern Europe the systems were closed and company based, although similar technology was used. Prefabrication factories were owned mainly by contractor companies. In Northern Europe open building systems were developed to guarantee compatibility between the products of different producers. At the beginning most prefabrication companies were owned by contractors, but some independent prefabrication producers later developed more competitive products and manufacturing technologies and entered the market. This helped further the development of open industrialisation and more flexible building systems. In the 1980s the technology of open industrialisation was applied effectively in the Netherlands. In Eastern Central Europe the industrialisation was widespread and national standard catalogue systems were applied. In some ways these systems were open, because several factories produced the catalogue products. The structural systems, products and even building designs were highly standardised owing to the use of several alternatives. Thus openness to users' needs was pure.

Once the mass production period of the 1970s was over, prefabricated building came into decline especially in Western Central Europe. However, in open industrialisation countries like Denmark, Finland, Norway and the Netherlands the share has since been stable or increasing. Systems and products have developed further towards increased performance quality and flexibility in design and use. The strict standardisation of products has mostly disappeared, and open system building has meant more systematic and compatible rules for design, production and changes under use. Besides building systems, information systems applied via computer aids, quality assurance systems and environmental systems are also being increasingly applied. Eastern Europe is seeing a transition phase in building technology and attempts towards better quality. Renewed interest in industrialisation will probably follow in the near future.

The start of this phase is marked by a decline in the share of prefabrication and an increase in traditional building methods like masonry.

The European approach to standardisation, product certification and free movement of goods, capital and people can, in the long run, activate co-operation for open industrialisation across national borders. However, to date standardisation has been highly conservative and does not support the acceptance of new types of products and production methods. In the future there is a potential to produce new standards which will be able to support new developments as well as a product certification system.

Below are some features from sample countries in Europe.

Denmark

The general trend is the increasing use of industrial products. Open industrialisation is an old tradition in Denmark. According to a study "The Innovation of the Building Process in Denmark" [9] the current trends in the building market is for both open and closed approaches. Steadily increasing industrial products are for closed systems. There is a low or even falling market for new buildings and steadily increasing market for renovations which use open systems. In the future maintenance will also demand open or closed systems.

There are trends in the demand for products and services which are for both open and closed systems. For total building there are regulations and diffuse market demands. Building products are open. Regulations are made for critical products and voluntary labelling is used on most products. General properties, workmanship and services on installations are changing from closed to open. In the future the focus will be on quality, indoor air standards, energy use, environmental issues and productivity. Closed consulting work will be constant and diffuse in the future [9].

Some trends can be seen for participants in the building process, which can be open or closed. Consultants are closed and contractors open or closed. The productivity of consultants and contractors is falling by 1% a year. New activities are emerging in quality- and environmental control, with increase of exports and specialisation. Producers are closed and productivity increasing by 1% a year. Export (40%) and increasing industrialisation is high [9].

There are also developing trends in open or closed controlling innovation. Basic research is open and publicly financed, but self-controlled. Applied research and building development are open, project oriented and public supported. Building development is controlled through public building projects. Product development is closed and financed and controlled by producers. Education and training are emphasising quality, new regulations and environmental matters, but not the building process and productivity [9].

A large panel system was developed in Denmark by Larsen & Nielsen. Unfortunately some of the details of this system, such as insulation, joints and connections, could not be utilised in other countries [30].

In a typical system of prefabrication Hojgaard and Schultz floor slabs are made of non-prestressed modular hollow core components 185 or 215 mm thick. They are supported on exterior sandwich wall panels and interior walls. Special sanitary wall elements with plumbing fixtures are also used. A system is composed of planar elements. The system could also serve as an example of a flexible system if examined independently of the general standardisation rules [52].

A three-dimensional box-system, Conbox, allows box units to be "stacked" into a cast on site or a prefabricated linear framing system. This removes them from the structural function and they can be used for any layout and height of building [52].

The closed system Unic 119 is a single-family house system. The building, composed of 37 prefabricated components, has been erected in many places in Denmark [52].

Larsen-Nielsen's LN-Nybo extends the idea of a closed system to residential buildings with varying apartment types, from one room (42 m^2) to six rooms (130 m^2). Different types of layouts can be created from the same group of components (exterior sandwich walls, interior bearing walls, and horizontal hollow core slabs) [52].

Finland

Industrialised building plays an important role in Finnish building construction. Its share is large - perhaps the largest in any market economy country. In multi-storey buildings the share of prefabrication is over 70% in apartment buildings and over 80% in office and commercial buildings. In industrial buildings the share of prefabrication is over 90%. In single family houses and detached houses the share of prefabrication is about 50%. Hollow core slabs represent the highest degree of industrialisation in precast concrete. In this business, the Finnish hollow core plant machinery and technology producers dominate the world market. Columns and beams are, for the most part, reinforced concrete, steel or wood. The production of steel and wood frames is highly automated while the production of concrete beams and slabs is carried out in pre-casting factories with semi-mechanised methods. Automation only applies to transportation and computers are used in the design and production planning. The exterior walls of multi-storey buildings are concrete sandwich elements, steel elements, lightweight steel sheet-thermal insulation elements or masonry structures. The exterior wall elements of detached and terraced houses are mainly of wood or concrete. The partition walls are mainly of gypsum and wood particle boards, but increasingly prefabricated modular system products are used. At the prefabrication of partitions highly mechanised and automated manufacturing processes are applied and the know-how and machinery of this manufacturing technology are exported all over the world.

Since the 1970s bathrooms and toilets have mostly been prefabricated modules. The same products are used on ships and sometimes exported. Currently there are some needs for improving the flexibility of these modules in order to offer greater possibilities for individual design and changes during use.

The development of industrialised building can be classified into generations [47]. The first industrialised generation was in the 1950s and 1960s and was based on the design and manufacture of individual elements. The structure of the elements was very similar to structures manufactured in-situ. Large panel systems were usually applied onto the frame. Traditional methods were used in the manufacture of these elements and the manufacturing took place mostly in precast concrete factories, but sometimes on site.

The second generation was that of the open element system, with the manufacture of compatible types of elements being mostly mechanised with pre-stressing techniques predominately used for load bearing structures. The second generation was developed for mass production in the late 1960s and 1970s. The flexibility of design was, therefore, subject to limitations which could only be overcome at fairly high additional costs. Some parts of the structural system, such as sandwich facades, beams and columns were and still are manufactured at quite low levels of mechanisation. The most developed industrial production techniques were applied to slab element production, with the use of extrusion and modern techniques of production planning and control.

The third generation, which developed in the late 1980s, aims to apply advanced industrial systems including hierarchical modular product methods, open computer aided design, production management and open, horizontally distributed production. The building concept is based on the effective and rapid site assembly of prefabricated components which can be produced in factories with flexible, mechanised, automated methods. The design of buildings and building components is based on computer assistance. Computer aided production planning and control throughout the entire building process is under development and will enable the effective use of modern methods of logistics and other production planning methods. Composite and mixed structures are increasingly used, e.g. steel frames together with concrete slabs in offices, commercial and industrial buildings. The increasing impact of modern user requirements stresses the need for compatibility between internal installations and structures.

'For open information systems and the application of computer aids in construction, a national development project called "RATAS" was started in 1984 [17], [35]. The RATAS project aims to produce national guidelines for open computer-integrated construction. As a first step an information infrastructure model, its data transfer rules and data transfer software tools were created. The BEC planning system was completed in 1987 making it possible to widely apply information technology for designing and producing precast concrete units as well as for realising whole construction projects. Guidelines for the partners in the construction process – architects, designers of structures and building services, contractors – as well as for use and maintenance have been published or are under preparation.

At the practical level a computer-integrated construction information system comprising the hardware, application programmes, databases and data transfer systems used in different organisations have been developed to conform to the standards and agreements necessary for integration. In Finland, guidelines for these kinds of solutions are laid down in certain development projects of the RATAS programme. The central idea is to standardise the systems of communication between different information systems without limiting the possibility of different parties to use and develop their preferred systems. Another key idea is the linking of information management to the design and planning systems. Design documents can be generated on the basis of the information needs specified by the user from plans and product models stored in a database [46].

The new Finnish building systems of the third generation, named TAT, comprise a hierarchical, modulated system containing functional and spatial systems, a structural system, systems for ventilation, water supply, electricity, information technology and waste management as well as the connecting dimensioning and tolerance system. The TAT system has five hierarchical levels: building, sub-building, module, component and basic element. The guidelines for design, production, service, product and method development are also defined. The TAT system provides designers with a common set of concepts, a dimensioning system, realisation principles for the technical systems, selection of component types, design instructions and model plans. Solutions at the highest hierarchical levels of the modulated system are free, and open to creative design. Module and component types are mainly modified products manufactured with flexible production methods, which can be adapted for a specific purpose to comply with a good plan. The most advanced in terms of standardisation are the highly technical components and connections. The following are the TAT targets:

- Increase in the freedom of design and quality/cost management.
- Industrialised and mechanised manufacture of components.
- The compatibility of technical systems.
- Easy assembly and installation of structural components.
- The optimisation and management of the design and production process.

There are several applications of the different system solutions, all of which are based on the TAT systems philosophy. The structural systems consist mainly of precast concrete, steel or composite structures and they are designed for housing, office buildings, commercial buildings and industrial buildings. The implementation of TAT systems has been done in different branches separately as individual product systems. Suppliers have developed deliveries at the following levels of hierarchy:

1. Component deliveries like beams, columns, slabs, facade units, partition wall units etc.
2. Assembly deliveries like bearing structural systems, facades, partitions, services etc.
3. Total deliveries, including assemblies and their design and installation, possibly even finishing.

The main principle of openness is that each delivery might be internally closed, but shall always be externally open, so that it can be joined to other deliveries both physically at the joints and in its performance.

In renovation more manual methods are applied on site together with industrial products. In the case of most old buildings the application of open systems is not possible. However, for the renovation of buildings from the mass production time of the 1960s and 1970s, renovation concepts applying open systems have been developed and their practical application has begun. Examples of such systems include "ENTRA" (by contractor NCC Puolimatka) and MOBIT (a group of 14 product suppliers). In both concepts the renovation process methods are based on continuous co-operation between contractor and suppliers in the framework of the open system. Besides the structural system, the compatible building service system also has an important role in the concept.

In the 1990s general application of ISO 9000 quality assurance systems has been done by all partners of the process. This has made the quality assurance system compatible, following the line of general openness of building systems. Right now the environmental quality systems based on ISO 14000 are at the start of their application, and will in the next few years be based on the open systems approach.

France

On-site construction dominates in France. Although component construction has long traditions, experiences have not been favourable [33] (Ansidéi) and it is considered more expensive than on-site building under present conditions [33] (Perälä). One reason is that open construction systems have not been developed since the earlier problem period, but firms have instead developed their own closed systems and co-operation between branches of industry is weak [33] (Salenius). Even today companies' closed systems dominate the market, although attempts and developments have been made and new open systems have been agreed upon and published by some groups of companies.

In spite of the progress made in many areas over the last few decades, building evolution is not comparable with the evolution of activity in other industrial branches. The traditional process reached its limits several years ago and is no longer proceeding. Production plants are at optimum productivity and operate using full automation, but the implementation process imposes the unavoidable fact that the site overrides the plant. Prefabrication with plants integrated into construction enterprises is totally unsuitable for the requirements that the new technical and economic rules now impose. In France, after many years of joint research between all the partners of the line –designers, design offices, manufacturers, contractors, administrations, etc. – a general dimensional convention has been defined for construction, which is probably the simplest and most dynamic which exists. Now that these rules exist, they must be used by designers so that downstream of the architect's project, other partners may produce their own services in total harmony and coherence with the works designed in this way [19] (Moulet).

The French sequential procedure is based on the organisation of work in the implementation phase, though it has expanded into versatile development of production. The fundamental idea is to improve productivity and cut waste in the process while minimising risks in general and creating the preconditions for effective utilisation of technology [32].

The French systems

Manufacturers, members of the FIB, and concrete shell work component producers have grouped together to harmonise their production. Their work enabled compatible component catalogues to come out in the early 1980s. The GCINQ association has created data processing tools for the use of the various building partners (developers, architects, contractors, engineering firms and manufacturers) [19] (Rhoul).

The Coignet system uses especially large floor slab elements (up to 4.50 m x 6.30 m) supported on interior walls 140 - 200 mm thick [52].

The French Opera system uses ribbed panels both for floor slabs and for exterior walls. The floor slabs of a total thickness of 280 mm are supported on concrete beams and columns. The exterior walls are completed on site with insulation and a decorative cover. All interior walls are non-load bearing [52].

The French Etoile system is one of the French school systems and uses planar floor slabs 210 mm thick, supported on steel columns, with prefabricated concrete footings. All interior and exterior walls are made of 160 mm thick plain concrete. The thermal insulation layer is connected to the inside surface of a facade in the course of the on-site finishing work [52].

The Ballot, Etoile and GBA systems are highly flexible open/closed systems. The systems use as their basic elements floor slabs supported on a grid of columns with large spans from 6.00 m to 7.20 m [52].

In 1982 the French Construction Plan instigated research into the creation and development of a computer program which could make their use easier and allow the development of overall industrialisation of the building industry. It is in this way that the Federation of the Concrete Industry FIB, the French Testing and Research Centre of the Concrete Industry CERIB and some professionals, as well as manufacturers of components had their proposal for research accepted by the French public authorities in the framework of the G5 Association. The G5 software now allows an architect's plan to be co-ordinated automatically and the industrial components for a project to be defined using a database of technical information for a particular manufacturer [26].

Thanks to G5 any architectural project can be carried out using industrial components without any pre-conditions being imposed on the designer and without obliging him to prepare his drawings according to the A.C.C. dimensional rules. At the level of constructing such a dimensionally rationalised project, the products are from the manufacturer's catalogue and can be manufactured using the same production process and tools [26].

Germany

In the mass production time of the 1960s many local and company-specific open-closed large panel systems were applied in Germany. Later the large panel systems were mainly changed into the application of composite precast and in-situ construction. Currently there are no general open building systems, but different component suppliers and contracting companies have their own building systems and concepts, where general components are applied in part. Some of the precast components are manufactured in highly automated factories. Typical of such systems are Filigran-type composite slab systems, where the precast slab serves as moulding for in situ concrete topping thus building a composite slab. This type of slab is produced annually at about 80 million m^2. The slabs are used together with beams or bearing walls as a structural system.

An example of such a system is the Miltz system, which is a linear system for industrial buildings [52]. The Milz Park system is a special closed system for car parking garages. It includes a small group of standard elements - frames, floor slabs, ramps, footings, exterior walls, and staircases. The system allows for a parking space of 2.60 m x 5.00 m per car and an access driveway of 6 m width [52].

In the new lands of Germany, previously the GDR, the experience gained in industrial construction formed the basis for the development of the housing construction series 70 (WBS 70). WBS 70 with a system depth of 12.0 m is a cross-wall construction method. The longitudinal exterior walls loaded via the longitudinal edge of the floor are used as stiffening for the building. On the basis of system segments of 6.0m x 12.0 m, but also 2.4 m and 3.6 m as well as 4.8 m x 12.0 m, a multitude of accommodation alternatives with 1 to 5 rooms were attained. The most important elements of the WBS 70 system are mainly of reinforced concrete and are the exterior wall elements, load-bearing interior wall elements and partition walls, floor elements, sanitary block units, stair and landing elements [22].

The notable construction methods for public buildings are characterised as follows [22]:

1. Precast wall construction method WBS 70 - storey height 3.30 m
2. Reinforced concrete skeleton construction method SKBM 72
3. Light multi-storey construction method Cottbus - LGBW
4. The reinforced-concrete prefab skeleton construction method SK Berlin DRS - an open building system utilising the composite concrete principle.

After the unification of Germany the new production of earlier building systems in the GDR virtually stopped, but renovation concepts for old building stock is now in progress. Industrialised concepts and product systems can be applied to this vast renovation activity also.

Generally the main trend in German building technology seems to be a slowly increasing share of industrially produced products, increasing automation in building product factories and further development of manufacturing equipment on site works. Some export companies especially from the Netherlands and Denmark are penetrating the German market with more industrialised production technology and products. This is somewhat difficult because of the strict standardisation and quality control systems, which do not support prefabrication and new types of products. Often a type approval is needed for new products.

The Netherlands

In the Dutch Open House principle the key idea is the distinction of 'levels': 'Support and Infill', fixed and flexible parts in housing. The dimensional grid co-ordination guarantees compatibility between support and infill. Dutch open building is mainly concerned with the customer's needs and changes in them, even during erection. The fundamental idea is to efficiently produce buildings and spaces of prefabricated components that meet customers' individual needs. The spaces must also be modifiable, which again emphasises life-cycle economics.

In general, the Netherlands has probably been the country in the European Community which (beside Finland and Denmark) has the greatest interest and shows the greatest concern for technical standardisation, conceptual innovation, 'industrialised' building and quality certification [33] (Mathurin). This is partly evidenced by the high level of prefabricated components used and the Netherlands seems to be one of the few European countries where the use of prefabricated components is among the highest in the western world [33] (Nurminen).

House building in the Netherlands is highly efficient, despite many traditional building methods. Precast concrete systems, applying mainly Finnish hollow core slab technology, together with precast concrete frames or walls within the dimensional grid modulation are in fact quite open system building, although no general open building system exists. Detached houses are increasingly built with hollow core slabs together with load bearing walls made of small scale lime sand elements. This open system is used increasingly in place of the traditional very stiff and closed in situ concrete tunnel mould system.

Several open infill systems are currently being developed. Four of these are:

1. The Matura Infill system, which is currently the most advanced system although its use is still new and very limited in scale.
2. Interlevel; this is a raised floor system on adjustable supports. Piping and cabling is done traditionally;
3. Espit; this system makes a raised floor by rolling out a mat with studs. The system comes with a dedicated partition wall system.
4. Actibo; in this system piping and cabling are fixed on the floor and then covered with ground aerated concrete, levelled and topped with floor boards 'floating' on the light weight concrete filling.

All the systems mentioned have one thing in common: they take away about 10 cm of the available floor to ceiling height. If any of these systems are adopted in refurbishment projects, the effect of raising the floor becomes apparent at the external doors of the unit. The systems differ in their level of detail, technical and organisational sophistication [14].

Spain

Open industrialisation and modular co-ordination is not generally perceived as advantageous in Spain. Until now, they have only favoured these systems for school building, recommending the use of modular increments, but limiting their use in practice to structural organisation and definition.

Generally speaking, the professional architect considers himself somewhat removed from this technical and cultural movement, and does not apply the principles of dimensional co-ordination as working tools, due to:

1. Lack of sufficient training and information about the modular co-ordination field.
2. A view that they are restrictive for his professional work without any countervailing technological or economic factors to justify such limitation or he may see them as mandatory performance standards imposed by the Government.
3. Their being associated with certain technological and political trends (e.g. heavy pre-casting) which have been poorly received both culturally and socially.
4. The general structure of development, dispersing the activities and giving power to urban renewal movements as alternatives to the production of extensive series of new buildings.

The industry is not capable of furnishing suitable components to these modular dimensions as alternatives to the conventional elements and methods [19] (Huete).

Sweden

Conventional building materials such as timber, cement, aggregate and others are currently utilised mostly in the form of manufactured articles and prefabricated building components in e.g. reinforced brick slabs, ready-mixed concrete, prestressed concrete columns and beams etc. [2].

Approximately 80% of dwellings in apartment buildings are built with bearing frames made of concrete. Facing materials are mostly facade blocks (50%), external timber panel walls (30%) and lightweight concrete (10%). Roofing materials are mostly roof-tiles (70%) or sheet metal (30%). Detached and semi-detached houses were mostly built with timber structures (over 80%), of which about 70% were built with prefabricated or partly prefabricated timber components. The facing materials mostly applied are timber (70%) and facade blocks (20%). Roofing is almost exclusively of tiles (98%) [2].

Office buildings are usually built with bearing frames of prefabricated concrete components. To some extent structural steel is utilised too. Prefabricated metal and concrete panel walls or lightweight concrete are usually utilised for facing, sheet metal for roofing. In the construction of industrial buildings prefabricated prestressed concrete components and to a certain extent structural steel are used for framing. Lightweight concrete and sheet metal are major facing and roofing materials [2].

General principles for dimensional co-ordination of prefabricated building components and building parts, the application rules for various frame systems and the corresponding tolerance rules are, in the main, accepted by the manufacturers of building materials, but not nearly to the same extent by the designers. The compatibility of prefabricated building components is not assured as there are no agreements on the performances and the geometry of joints [2].

Currently applied systems of prefabricated building components and building parts are in some quarters considered as open systems. Prefabricated components for frames and envelopes from various manufacturers can certainly be made compatible - after costly adjustments either in the factories or on the building site. This is considered one of the reasons why during the 1990s the share of "semi-prefabricated" composites (prefab + in situ) in the frames of apartment buildings grew almost to double the share of fully prefabricated components [2].

As far as the building contractors and manufacturers of prefabricated building components are concerned, computer aided design and manufacturing is an internal organisational task shielded from potential users such as the clients and/or their consultants. The lack of systematic and open technical information contributes to the hesitation of the designers and the clients to use prefabricated building components. Thus the lion's share of the components are designed by the manufacturers. There is, however, a somewhat greater openness in communication between clients and their consultants, but this openness is not particularly dependent on computer-based routines [2].

Planning of the building's production is an internal affair of the contractors. Large and medium sized construction firms are the predominant owners of the building systems on the market. They nominally compete with each other by offering some isolated products, but there are no reasonable incentives for a more extensive exchange of the various prefabricated products between the systems. The organisation of the construction process and any detailed account of its environmental performances are shielded from the external interference of the clients and/or from their consultants. The applied building techniques' adaptability to the requirements specified for a particular site is ultimately disclosed in the form of "the product" - which is a ready-made building offered to the client and made using the contractors' own building system [2].

According to a state of the art survey 'Industrialised building for apartment buildings with open building systems' (dimensional co-ordination, tolerances, compatible components and joints), the market share is 0 - 40% depending on the part of the building. The market share of full prefabrication for all parts is 0 - 20% and increasing. The share of composites is about 20 - 40%. For buildings where open computer aided design systems, modern methods of production planning and control, systematic quality assurance methods and advanced modular production infrastructure and organisation (construction with effective use of external deliveries and services) are used the share is 0 - 20% and increasing for all parts of building. For bearing frame structures where quality assurance systems and advanced modular production methods have been used the share is 20 - 40% and increasing. For office buildings and single family houses and detached buildings the shares are about 20 - 40% and increasing. For single family houses and detached buildings the bearing frame and facade are 60 - 80% fully prefabricated. Generally speaking the share of the buildings where modern industrial principles have been used is increasing [4].

United Kingdom

We have seen how there are sharp boundaries in Britain between most of the existing building systems and that these boundaries are identified both technologically and administratively [45].

Atcost is a linear system for industrial buildings [52].

Bison system is a flexible (open-closed) system [52].

A British Spanclad system is typical for large, heavily loaded public and industrial buildings. The system employs Double Tee components of varying height supported on beams, columns, and exterior walls can also be made of Double Tee components [52].

CLASP, which is nearly 40 years old, is the most successful - possibly the only successful - British building system. The system is still alive, much diminished in scale, but in good health and based in Nottingham. CLASP is a fully developed, mature system that is turning out to be remarkably flexible. The system was originally designed to cope with mining subsidence by moving with the forces of nature, Japanese house-fashion, rather than opposing them. The system's great advantages were seen to be in the way it enabled good-quality schools to be built within cost yardsticks, and allowed speedy erection by relatively unskilled labour, at a time of great demand in construction, and the beginning of the widespread de-skilling of the industry's workforce [36].

Currently, there are attempts to apply increasingly prefabrication especially in office, commercial and industrial buildings. Composite construction methods are popular at the newest developments [54].

Norway

In Norway prefabrication has developed rapidly, especially during the 1980s. The precast concrete industry has been active in developing open structural systems together with their design and production. The production infrastructure, technology and products are quite similar to those in Denmark, Finland and the Netherlands. Also prefabricated wooden building systems have been developed and exported to the EU, especially to Germany. The share of open industrialisation is on the same high level as in Finland and the Netherlands. The precast concrete producers are serving the contractors with component deliveries, assembly deliveries and total deliveries.

Switzerland

In Switzerland building is quite a local activity and carried out mainly with traditional site methods, but using industrially produced components and products. There are no open systems and the prefabricated systems currently in use are closed company systems. Often automation is at a very advanced level in the prefabrication factory.

The Peikert housing system is a flexible (open/closed) system. The system uses as its basic design unit a multi-module of 9 M (M = 100 mm) and a set of flexible components based on this module. The thickness of a bearing wall is 150 mm, while the thickness of the floor is 180 mm [52].

In the Variel three-dimensional box-system, the units include basically only floor, roof slabs, and concrete corner posts. It leaves the designer and the user the highest flexibility of space utilisation, but may require a special set up in the plant for completing the finishing works [52].

Czech and Slovakian Republics

In the Czech and Slovakian Republics the system building was earlier similar to systems in other Eastern European countries. Quite similar systems to catalogue systems, e.g. MS,S, INTEGRO and P1 have been used [52]. In the new market economy the technology will be changed and the share of renovation will increase. The current state and future visions seem to be quite similar to those in the new lands of Germany.

System building has strongly declined and more conventional building methods now dominate. The renovation and modernisation of old building stock is being prioritised. In the new market economy the technology will be changed and the share of renovation will increase.

Croatia, Slovenia and Yugoslavia

The Yugoslavian IMS system has precast columns, one, two, or three storeys high, spaced at 3 - 9 m spans (4.20 m x 4.20 m grid is most common for residential buildings). Large ribbed slabs, 220 - 360 mm thick, are attached to columns by post-stressing the whole floor after its erection. Sandwich exterior walls, stairs, and partitions are added to the structure after its erection. Elements 150 mm thick are used both as inter-dwelling partitions and shear walls to provide sufficient rigidity to the system [52].

Another Yugoslavian system - Sistem 50 - uses as its basic floor component a cassette floor of 3.5 m x 3.5 m. Four slab components, connected between them to columns by poststressing, create a free area between columns of about 50 m^2. Sistem 50 is a more flexible open-closed system [52].

The Yugoslav construction industry has achieved its remarkable results in the course of the past decades both in the country and abroad. Times of crisis, which came as a result of the recent war, have led to a smaller market for construction, inefficient running of construction businesses, badly organised processes and low productivity. In such conditions the construction industry, therefore residential construction as well, are unable to follow international trends. However, the return to traditional construction, in the desire to give employment to the workers at all costs, is not going to lead to more efficient business [55].

The development of the industrialisation of building construction, residential as well as for other purposes, will be the direction of development for the next century, in the course of which the stress will be placed on construction quality and economy. The industrialisation will not be its own purpose. Instead, the industrial product will be a means for improving the quality of the building as well as the economic efficiency of its construction. Judging by the world-wide development of building, residential construction systems of the large type, either large panel or skeleton systems, are losing the race to flexible systems made of blocks or smaller elements, which are more adaptable to the customers requirements. It is necessary to come up with new ways of development and adaptation using valuable past experiences and knowledge, all in the direction of world-wide tendencies [55].

Hungary

The catalogue systems have earlier been applied also in Hungary. These include:

-   UNIVAZ, a linear system for public buildings [52].
-   Univaz and Elvaz, more flexible open-closed systems.

The range of applicability of closed systems could be extended by introduction of "cell" (or "block") elements. A cell is referred to here as a design element of room-size dimensions, which can be used as a building block for different layouts. Such cells have been used in Hungary by a prefabricating factory, as typified elements of kitchens, bathrooms, living rooms, staircases, and so on. These cells could be viewed as a mirror image of the Pazner-Brandt block concept. They were produced from standard elements but could be varied by an architect in different ways to attain distinctive layouts [52].

Also in Hungary the building technology is undergoing rapid change and the share of renovation is increasing.

Poland

In Poland the predominant building system is the large panel system. An example of a partly open, partly closed system is the Polish system W-70. The system uses a design grid of 6M x 6M (M = 100 mm) and four basic floor spans: 2.40 m, 3.60 m, 4.80 m and 6.00 m, supported on interior bearing walls. The preferred spans were established following an analysis of possible residential room sizes which they could economically accommodate (a similar functional analysis of economic span widths between supports). The system uses interior and exterior wall units ranging between 1.20 m and 6.00 m in length with various openings, and special elements for heat units, stairs, roofs, and basements. All these components are arranged in catalogues with typical connection details. An architect using the system has a practically unlimited choice of building layouts, provided that he or she uses a modular grid of 6M x 6M and provides load-bearing wall (interior and exterior) spacing as required relevant to the floor spans that have been used [52].

A flexible closed cell system is available based on "empty" cells, that is, modular room-size area units using standard components for their construction. Such cells were used in the Polish system Szczecin (2.40 m x 4.80 m) and OWT-67 (2.70 m x 4.80 m) for construction of various apartment layouts [52].

Romania [18]

The newest construction system uses a single main type of space prefab-unit by means of which are erected multi-storey frame structures without apparent beam floors. The system describes the technology of unit fabrication, of mounting and assembling the structure, and the consumption parameters of materials recorded upon achievement of some three-storey administrative buildings. The construction system presents some important advantages: one single main type of unit, high speed of mounting, the interior free division facilitated by the lack of interior beams, reduced material consumption and manpower, as well as the achievement, with the same type of unit, of a large variety of constructions: blocks of flats, hotels, administrative buildings, schools and research institutions.

The S.F.U. system conceived and achieved in Romania uses only a single main type of prefab space unit with self-stability at mounting by means of which frame storied structures are made without apparent beams for floors, as well as a high mounting speed and reduced consumption of materials and manpower.

In the S.F.U. system the structure of a building is achieved by an adjoining arrangement in a horizontal and vertical plane of some prefab units in the form of space frames, and the continuity of the transversal beams by steel loops, of the longitudinal ones by pre-stressing and of the columns by welding or hinges. The uniqueness of the S.F.U. system lies in the shape of the space prefab unit, in adjoining way of arrangement of prefab units and in the way of joining by steel loops and prestressing, which gives the structures a high safety and stability and makes them suitable in seismic zones.

The fabrication technology of the space units is simple, these being cast in place in closed halls or open platforms provided with a bridge crane. It is recommended that the formwork for the inferior columns be arranged in cavities so that the slabs of the units are cast directly on the mosaic platforms. The units are manufactured in three technological cycles each lasting 8 hours. In the first cycle, prefab units made the previous day are struck, followed by platform clearing and mounting of formworks and the reinforcement. In the second cycle, concrete casting is done by a special carrier-distributor. The third cycle comprises the heat treatment.

Transport to the mounting site is done on trucks fitted with steel beams on which the units are laid uniformly with the floor slab. Mounting is very simple and is carried out with autocranes or tower cranes with a capacity of 5 tonne/metres, directly from the trucks or from a temporary storage area. It involves adjoining the units horizontally and then simply superimposing them. Mounting may be continued without joining the beams and columns, since the space units have a proper resistance and stability at mounting which gives this operation its high speed.

Assembly of the transversal beams and ribs is done with steel loops without welding, that of the longitudinal beams by prestressing with two ropes placed in two channels formed during unit prefab, and the columns are hinge-assembled with steel bolts, or continued by reinforcement welding. The end walls and the interior division walls are made by known solutions and methods.

Israel

The main building material in Israel is reinforced concrete. Three main technologies are employed in building construction – the traditional technology, industrialised monolithic construction and industrialised prefabricated construction. About 60 - 70% of residential buildings are multi-family houses 3 - 12 storeys high, the rest being 1 - 2 family low-rise houses [51], [56].

The prevailing "traditional" technology employs framing of cast in situ reinforced concrete columns and flat slabs, and exterior walls and partitions made of hollow concrete block masonry. Two layers of plaster are used for interior wall finish and three layers of stucco for exterior wall finish. Terrazzo tiles are used for floor finish [51].

Roughly 20 - 30% of building construction (overall) employs industrialised methods. The industrial monolithic construction uses tunnel forms for cast in situ cross-walls and floor slabs. The exterior walls are made of prefabricated elements or of block masonry [51].

The prefabricated construction uses prefabricated bearing walls or columns, prefabricated floor slabs and prefabricated exterior walls. Hollow core slab and T-beam floors are used in commercial and industrial buildings. Room size solid slabs are mostly used in residential construction, as tenants do not wish to see joints in their ceilings. Gypsum board (dry-wall) partitions are mostly used in all types of industrialised construction [51].

Small 1 - 2 storey residential buildings are mostly built with traditional methods, although there is an increasing use (if not yet significant) of lightweight construction methods based on building frames of steel or timber [51].

There are at present close to 20 plants which produce concrete building components. Prefabricated elements with the largest demand are hollow-core prestressed slabs used in non-residential buildings. The residential prefabricated buildings use room-size, solid-slab components. Exterior, prefabricated walls are used both in residential and in non-residential construction. The exterior walls consist of four layers: two layers of concrete, an insulation layer of polystyrene between them, and in decorative finishes granolitic or decorative stone. Prefabricated T-beams and bridge beams of standard profiles and sizes are very frequently used in bridges. Concrete prefabricated columns are used less frequently [51].

Industrialised installation modules are used in commercial and industrial buildings and occasionally in multistorey residential buildings [51].

Open industrialisation in its full sense, i.e. full compatility of dimensions, tolerances and joints, is almost non-existent in Israel. A modular co-ordination building code (IS 417) was introduced in the 1790s and defines modular dimensions and tolerances. It was implemented in government assisted social building. Today these buildings constitute only a small proportion of the total output and the code is rarely implemented in practice [51]. Prefabrication is amply covered by the Israeli codes and specifications. The production and assembly of precast elements is also covered by Israeli general specifications. Industrialised building is taught at the Technion — Israel Institute of Technology at both undergraduate and the graduate levels, but only as selective subjects. The main emphasis at graduate level is on managerial aspects and system approach. An acute shortage of construction labour has forced the Israeli Ministry of Building Construction to support an extensive research programme for automation of construction activities. The programme is run by the National Building Research Institute of Technion, and its objectives have been defined as follows:

- To develop a concept for automated building construction.
- To examine within this concept the main technologies which will be necessary for implementation of this concept.

- To examine the economic implications of employment of the concept.

A robot, named TAMIR, has been developed to execute the following tasks: painting, plastering, partition building and tiling [56].

The closest construction in Israel gets to an open system is utilisation of prefabricated hollow core slabs of standard widths of 1.20 m and 0.9 m (12M and 9M respectively) and lengths in increments of 0.30 m (3M). Modular stair elements are also in use. These elements are used in conjunction with both industrialised and traditional construction [51].

A typical field plant Israeli system which is inexpensive and is used by most Israeli prefabrication companies comprises 120 - 130 mm thick room-size floor slabs, 150 mm thick room-size floor slabs, 150 mm thick load-bearing cross walls, 200 - 250 mm thick exterior walls, and 70 mm thick partitions [52].

The Israeli Yuval Gad system uses 7.00 m x 2.00 m cassette slabs as the main floor element. The slabs are supported on interior walls and diamond-shaped exterior wall panels [52].

Two storey "cottages" of the Mabat system have been built in Israel. The project is a combination of box units, which constitute the main building element, and various planar elements — for example gables and partitions [52].

The following trends are expected to greatly affect in future the development of industrialisation in Israel [56]:

1. Economic growth.
2. The shortage of construction labour.
3. A growing demand for diversification in residential and non-residential building.
4. A growing demand for quality and aesthetics.
5. Growing dependence on computer aided design.

Summing up, it seems that the prevailing trends in economy, technology and customer preference definitely favour a growing implementation of prefabricated methods in Israel and their further development towards increased extent of automation. The implementation of these methods can be aided to a large extent, in Israel as in many other countries, by an extensive effort in several major areas:

1. A comprehensive education of engineers, architects, and contractors in the particular benefits, nature, and special requirements of prefabricated construction.

2. A marketing effort which will emphasise the benefits of industrialised buildings and focus on areas where these methods have a definite advantage over traditional building methods.
3. An extensive standardisation of common elements — slabs, walls, partitions, facades, beams, columns, produced by different plants — their shape, range of dimensions, joints, production specifications and performance.
4. Further technological development oriented towards improvement of industrialised methods and introduction of automation [56].

3. 2. 2 Russia and GIS countries

Russia

The development of mass-scale housing construction, on the grounds of typisation and standardisation, has been closely connected with its industrialisation, with the organisation of a special base for home building, and with the transfer of production of structures for residential buildings to be planned. Large-panel home building (LPH) has advanced rapidly since the end of the 1950s, its share having grown recently to 52% of the overall amount of state and co-operative construction. In urban housing construction, its share is 60 - 62%. In Moscow, Leningrad, Kiev, Gorky and other large cities, the share is 75% and above. In the early 1970s, along with LPH other types of industrialised home building began to develop such as modules, in situ concrete, and in small townships and villages fully prefabricated timber. However, these are not widely used [21].

Industrialisation has changed the type of building materials as well. Previously, bricks and other small piece materials accounted for one-third of all state and co-operative construction. Now, concrete and reinforced concrete prefabricated structures, foundation blocks and piles, cellular slabs, staircases, etc. are being used on a wide scale [21].

The large panel building system currently in use was developed in France by Raymond Camus and Eduard Coignet in the 1950s. From France, the system was exported to other European countries. In the last 30 years, many millions of apartments have been constructed using adaptations of this system, especially in the former Soviet Union and Eastern European countries [30].

Housing, office, commercial and industrial buildings, which are produced within investments of foreign companies and other organisations, are often built by foreign companies applying their building concepts, which are often quite industrialised techniques using local materials together with imported industrial components. Traditional masonry building technique has gained an increased share in the building of one-family and detached houses.

Russia's current transition phase towards a market economy has created growing interest in open industrialisation. The objectives are the increase of architectural and technical quality, flexibility in design and use and energy efficiency. Open industrialisation will be one possibility in the future development of Russian building technology, but no clear steps in that direction have been reported to date.

3.2.3 Northern Asia

Japan

Open building has been developed in Japan for more than 20 years in both the public and private sectors. While the concept has reached a threshold at which significant advances are now possible, a number of problems remain to be solved [29].

One family houses have long been built using wooden post and beam construction. This conventional construction method for detached houses is not industrialised but well systematised using a modular grid system. It can be called an open building system. Some companies started to supply industrialised (prefabricated) houses in the 1960s. These are closed systems, but most of them also use a modular grid system. In Japan 20% of total annual house production (700,000) is prefabricated industrialised houses [23].

Apartment buildings are built using reinforced concrete construction. Most of them are made with concrete cast in situ. Some private condominiums are also constructed using precast concrete, but they are neither well systematised nor dimensionally co-ordinated. Only some public apartment houses have been built based on a co-ordinated modular system. The construction of structures is not well industrialised, but some of the components for apartment buildings are produced in an industrialised way. Bathroom units are produced as a three dimensional component by many companies [23].

Office buildings are built using reinforced concrete or steel construction. Low-rise or middle-rise buildings are not industrialised, but high rise office buildings are built using steel construction and designed on a modular grid system. External walls of such buildings are supplied by curtain wall manufacturers, and internal partitions are made by specialised companies. Systematised ceilings from component manufacturers are used in the high rise buildings [23].

Different materials are used in different types of buildings. Most detached houses are built using wood. Steel is used for the structural members in industrialised houses. Precast concrete is also used for prefabricated houses, but the number of detached houses using precast concrete is not large. Middle-rise and high-rise apartment buildings and condominiums are built with reinforced concrete or hybrid combinations of steel and reinforced concrete. Low-rise apartments are built using wood or steel. Most of them have very small dwelling units. Office buildings are mainly built with reinforced concrete. Some high-rise buildings and skyscrapers are steel constructions with curtain walls.

For detached houses, prefabricated structural systems are supplied as products of industrialised house manufacturers. Three-dimensional space units for detached houses and low-rise apartment houses are also produced in fully automated factories. They are good examples of successful closed systems. Conventional wooden houses have not been industrialised, but were divided into some systematic subsystems such as the "tatami" floor system and "fusuma" movable partition system, both of which were well co-ordinated. The joints of posts and beams of conventional wooden houses were worked out by carpenters, and are assembled into the shape of the house on site in just one day.

Industrialised manufacturing systems are being applied to conventional wooden houses. This is called the 'pre-cut' system. Recently a Cad-Cam system for designing and making wooden members has been developed. The wooden posts and beams are factory made with the right joints the day after the plan of the house is fixed. Open components for detached houses such as aluminium window sashes and kitchen equipment are supplied from other manufacturers. The bathroom unit is a typical industrialised component from the open market. Recently, for apartment buildings the combination of precast concrete and cast in-situ concrete has become fashionable. Other components for dwelling units such as partitions are not systematised. For office building, industrialised components such as partitions, ceiling system, free access floor system, toilet equipment system etc. are supplied by a range of manufacturers [23].

Japan is a country where the "infill" industry, namely making building components such as kitchen systems and bathroom units, is highly developed. This development, in a sense, is the result of co-operation among building industries encouraged by the Ministry of Construction and Ministry of International Trade and Industry. [46] After the experimental projects, studies of two important housing system were carried out during the 1980s which form the basis for future open housing developments in Japan.

- Century Housing System is a classification of building parts in accordance with their life span.
- Two-step Housing Supply System: a study of building deregulation to enable the separation of supplying support and that of infill [47].

The efforts made in Japan to develop the support and infill concept, and technologies and techniques to manage open housing design and production can be summarised as follows:

- Technology and industrial capabilities have accumulated in general contractor and prefab companies.
- The expression of the three level design method, as proposed by the SAR, which is to determine infill, support and urban tissue based on agreements in families, neighbourhoods and communities, has not yet been fully realised except in a few urban renewal projects.
- Support and infill technologies are developed in large general contractor and prefabricated housing companies, and are only for house production activities for the market segment in which each corporation deals. These systems should be called closed since they are open only within each corporation.
- This type of support and infill housing was developed in Japan for the following reasons:

  1. The volume of production for each corporation is large enough to enjoy the benefits of systematisation.
  2. Computerised manufacturing can supply products of great variety with high efficiency [49].

The Japanese Misawa system combines steel box units with cast in plant exterior light-weight concrete walls [52].

Korea

The Korea National Housing Corporation's Housing Research Institute produced a proposal "Principles and Recommendation for the Modular Co-ordination building in Korea" in 1996. The proposal aimed to develop National Standards for fostering modular co-ordination in Korea [42].

China [27] (Goupan)

Currently, stress is given to the improvement and extension of application of existing industrialised building systems. In the meantime, a technological structure at multi-levels of mechanisation and semi-mechanisation combined with manual operation should cope with the actual conditions and be appropriately adopted, so as to make every effort to save investment, reduce expenses, improve the engineering quality and accelerate construction.

A housing system with small prestressed concrete members and concrete hollow blocks uses small prestressed concrete members for the floor and roof, and concrete hollow blocks for the wall. The construction is easy and light-weight. The small hollow blocks can be easily moved and handled. Prestressed beams, purlins and T-shaped slabs can be erected with simple tools. This load bearing wall construction can be used for buildings as tall as six storeys in regions of seismic activity within the intensity of 7 degrees. The cost is comparatively low and the system is suitable for housing and medium sized public buildings.

Medium-size block and precast slab system is a housing system using blocks made of concrete for the walls and precast reinforced concrete hollow slabs for the floors and roof. The blocks have been developed into four different types with an average weight of about 135 kg per piece. The construction equipment used is generally light, simple, mobile and easily handled. Light-duty tower cranes or hoists are used for vertical transportation, while the machine for laying and erection is mainly light-duty derricks. The system is mainly used for the construction of multi-storey residences.

The "Cast-in-situ Concrete Inside and Brick Outside" Building System is a system with internal walls of reinforced concrete cast in situ using large steel formwork. The external walls are of brick and floor slabs, stairs etc. use prefabricated products. The system achieves good integration of masonry, cast-in-situ concrete and prefabrication techniques into one monolithic structure resistant to earthquakes. It has merits of heat insulation, water-proofing, sound insulation and aesthetic finishing and its cost is comparatively low as well. It has been used widely to build a number of four to six storey residences in seismically active regions.

The Prefabricated Large Panel System is a building system with prefabricated room-sized reinforced concrete panels for walls, floors and roofs. The finished skin of external wall-panels is also prefabricated in the plant. The components are large, about 5 tonnes per piece, which necessitates using large tower cranes and a skilled crew. More capital investment is needed for the establishment of a prefabrication plant. The system has been used for multi-storey residential buildings. Its competitive ability is rather limited and therefore, restricted to the development of medium to high rise (12-16 storey) buildings only.

The Frame and Lightweight Wall Building is a system with a load-bearing structure of frame and floor slab type, or frame, shearing wall and floor slab type. While the columns, beams and slabs may all be prefabricated elements, or the beams and slabs prefabricated, the shear walls and columns are cast in situ using steel formwork. This system with characteristics of light weight (about 500 - 800 kg/m^2 of floor area) is favoured for its seismic resistance. The cost is rather prohibitive.

The Slip-Form System is a building system with reinforced concrete walls cast by slip-form and the floor is also cast in situ storey by storey. It makes the structure more monolithic and integrative and it is possible to meet a variety of building plans. This approach is mainly used for high-rise housing and offices to provide a seismic resistant structure of tube and shear wall.

The cast-in-situ combined with prefabrication is a building system using large-sized steel formwork to cast reinforced concrete internal walls on site with prefabricated large panels for external walls and room-sized slabs for floors and roofs. Since the system has the characteristics of better integrity, stronger seismic resistance, faster speed of construction, comparatively lower cost and the simplicity of equipment with a process which is easier to master as well as a less capital investment, it has developed quickly in certain large cities.

The development of industrialised building systems in China is mainly focusing on urgently solving the shortage of housing. Along with the development of the national economy and the relaxation of housing supplies the agenda has been slowly changing to the key problems of the improvement of residential quality, improving the flexibility of layout and partitioning internally within the dwelling unit, as well as the possibility of enlarging the space in a dwelling unit where better accommodation might be needed in the future.

3. 2. 4 South and Southeast Asia

Afghanistan [27] (Faruq)

In general the local construction materials in Afghanistan are stone, mud, mud bricks, fired bricks, sand and gravel, lime stone, gypsum, cement, timber (different sizes and kinds) and straw. All the rest of the constructional materials such as glass, hardware, steel, sheet metal, toilet fixtures, pipes, electrical materials and sanitation materials are imported.

In urban areas, a combination of local and imported materials is used applying local and advanced technologies. Besides the commonly used materials for flooring and roofing, concrete hollow blocks are being introduced.

At present, the use of hollow blocks due to its efficiency in time, cost, and materials has replaced bricks in residential and other civic buildings to about 10% per year. Besides that, hollow blocks have been used in roofs along with reinforced concrete prefabricated plates. Hollow blocks are produced in various sizes and used for walls, partitions, fences, compound walls and as a filling material in frame structures.

Roofs are constructed with cloths, straw, mud and mud bricks in the rural areas and with mud, fired bricks, reinforced cement concrete, hollow blocks and sheet metal in urban areas. Therefore, hollow blocks, as a substitute for bricks and monolithic concrete, has been readily accepted by designers and people in almost all buildings. The production of hollow blocks is done both in open air and in temporary shelters.

The Housing Construction Unit Kabul, which functions under the direct control of Kabul's Municipal Corporation, produces all elements for construction of pre-fabricated housing blocks. These blocks consist of reinforced concrete walls, roofs and floors, wooden doors and windows. Each block contains about 60 apartments and is constructed on five floors. At present the production capacity has gone up to 880 apartments per year, and with the completion of the full factory its production capacity will be doubled within one or two years.

Bangladesh [27] (Roy)

In Bangladesh natural resources of conventional building materials like steel, stone, timber, lime-stone, etc. cannot cope with the increasing demand. The local building materials are bamboo, mud and bricks. The houses which the majority of rural and urban dwellers inhabit are built mainly with bamboo and straw for walls and roofs. Cement is used as a binding material in construction work of all reinforced concrete and brick masonry structures, plastering work for water tightness, decorative, ornamental and pointing work. The country has two cement producing units. Only 5 - 10% cement is produced with local limestone. Gravel is used as coarse aggregate in concrete.

The country is short of natural stones. Burnt clay bricks are used in place of stones and are the only building materials versatile in its use. In Bangladesh, around 2000 brick industries are producing bricks with only 5 - 10 industries semi-mechanised or mechanised. Brick is available in almost all parts of the country. Most of the bricks are manufactured during the dry season, so production is mainly seasonal.

Bangladesh has no natural deposit of iron, so steel is imported either in the form of raw materials or finished product. Timber is used for walling, roofing, flooring, door, window frame and shutter. Bamboo is the most common building material in rural Bangladesh. About 46% of houses in Bangladesh are built with straw and bamboo.

Concrete blocks are made by mixing Portland cement, water and aggregate which may be sand, gravel, crushed stone or crushed old concrete depending upon the weight and texture. In Bangladesh, there is no factory for producing building elements like concrete hollow blocks. The concrete block housing is, therefore, negligible.

The present day construction system is cast in-situ concrete, of up to five storeys constructed with 25 mm thick load bearing wall on spread foundations. All elements are constructed on site. The following materials are used for wall and ceiling construction: mud, brick, concrete block, corrugated iron sheet, bamboo, timber, plywood, asbestos, cement etc. The different types of floors which are commonly used for ground floor construction are mud, brick, concrete, timber, terrazzo, mosaic and pre-cast concrete. The following roofing methods are flat roof, pitched roof, with coverings of thatch, concrete roof tiles and asbestos cement sheet

The Housing and Building Research Institute has already experimented with a number of techniques for constructing structural floors and roofs by pre-cast concrete elements with a view to lowering the cost. Partial pre-fabrication has resulted in savings of cement, steel and cost. Timber for temporary centering and shuttering works has been virtually eliminated.

Sri Lanka [27] (Ediriweera)

Concrete is the most widely used construction material in Sri Lanka. The use of structural steel has fallen considerably mainly due to its high import cost. However, in certain instances, steel is preferred to concrete or timber due to ease of construction and for architectural reasons. Occasionally, steel columns, beams and trusses are used in the construction of warehouses, pavilions and single storey buildings. In certain other instances, steel is used in the construction of grandstands to expedite their construction.

A few decades ago timber was used extensively in the construction industry. However, due to its scarcity, timber is mostly used for door and window frames and sashes. Until recently, wooden formwork had practically been the only material for making formwork. Timber is used for structural elements only in single storey dwelling units. During the last $1^{1}/_{2}$ to 2 years, timber has been once again used for structural elements, especially roof trusses, with the introduction of Gang-Nail timber trusses in Sri Lanka.

Most used are pre-fabricated building components like columns, beams, slabs and purlins. In this method the building is seen as a structural skeleton to carry and transfer the loads, and as a non-load bearing infill consisting of walls and roof cladding to provide shelter and security. Depending on the user's requirement, architectural appearance and the time available for construction, one or several pre-fabricated units are used. The most popular method for single storey structures is the use of the precast building frame.

The building frame is widely used in Sri-Lanka in the construction of low-cost houses, school buildings and warehouses, because it provides a comparatively cheap structure for these buildings, which is a simple, yet sturdy, permanent structure which can be erected within a short time. The building frame consists of precast column pockets, columns, beams, purlins and brackets and bolts necessary for fixing. Precast members are cast in a yard and transported to the site. First, the column pockets are placed in excavated pits and the columns are lowered into the column pockets and positioned.

Currently, there is a heavy demand for these building frames, as they could be erected even by non-technical staff following the manufacturer's instructions. Another attraction of this system is the facility for extensions to the existing building. This involves the simple addition of a couple of frames as required. Although it is cheap and easy to erect these buildings, architecturally they are not pleasing. Therefore, this type is generally used for classroom blocks in primary and secondary schools, warehouses, temporary buildings and low cost houses. Due to the high capital investment involved, the use of slip form shutters and sophisticated machinery is restricted to large construction sites, where foreign contractors are engaged in heavy construction.

Precast components are also used in multi-storeyed building construction. Depending on the nature and the type of building, one or more components may be used. There are several multi-storeyed office buildings constructed in Colombo, the capital, using pre-stressed precast beams. A series of these beams, spanning between either pre-stressed main beams or in situ main beams, constitutes the floor. A common feature of all these office buildings, where pre-cast floor beams are used, is the considerably large spans required for architectural reasons and specific user requirements such as auditoriums, restaurants, car parks etc. In certain instances the clear spans are as large as 45' (14 Metres). These floor beams are also extensively used in the construction of grandstands in stadia. Generally, all architects prefer to have large spans in grandstands to reduce the number of columns, which if positioned awkwardly hinder a clear vision of the activities in the ground. In addition to this, the construction of in situ tiers are expensive and time consuming. Further, as most of these floor beams are slender, they add to the beauty of the building.

Pre-cast floor slabs were extensively used several years ago in the construction of low cost housing schemes. Use of these floor slabs was convenient as load-bearing walls in most housing units were arranged in a regular pattern and at close intervals i.e. at about 3 metre centres. Another popular technology is the use of cement and sand hollow blocks for load bearing walls.

Although it is considerably cheap and less time-consuming to construct with pre-cast components, the use of these pre-cast components is restricted to large scale construction works. This is mainly due to the expertise needed in casting, transporting, handling and erection of the units. The heavy weight of the components prevents manual handling. In Sri Lanka, there are very few contractors who have the technical expertise and the heavy machinery needed for transportation, handling and erection of the units. Unfortunately, this has resulted in a limited use of pre-cast elements.

Another area where pre-cast elements are extensively used is in the construction of highway bridges and pedestrian bridges. In these two instances, the nature of construction itself restricts the use of in situ concrete.

Thailand [27] (Chaimungkalanon)

Thailand is self-sufficient in building materials, producing 7.9 million tons of cement in 1986 while total consumption varies around 6-7 million tons per year.

Wood was the most popular building material in the past. Until recently, wooden formwork was practically only for reinforced concrete structural and cast in place construction.

In 1962, a private construction company developed a system called Seacon. This system is suitable for a country where unskilled labour is plentiful and cheap. There is no heavy machinery or high technology involved. The system is applicable to almost all types of buildings. For a standard house that normally takes 6 to 8 months to construct, this system can reduce the construction period to 1.5 to 3 months. The construction cost can also be reduced by 10% to 20% (depending upon the efficiency of the management). Seacon is a skeleton system with columns and beams. The erection of pre-fabricated components begins with the setting up of latticed steel posts on the footing and welding them to the steel reinforcement projecting from it. Beams and wall panels are then lifted and placed between the steel posts. The remaining work such as roofing, ceilings, partition, plumbing, electrical work and painting, is carried out in much the same manner as in conventional construction. Until now more than one million square metres of housing, shop-houses, shopping centres and flats etc. have been constructed using this system. In one way or another this system has great influence on many other systems that have followed. Further developments have been brought about by the Seacon system: Another private company developed a system based on Seacon's concept; the difference being that all the reinforced concrete components are pre-fabricated in a factory. This system was also applied to the building of many private housing projects in Bangkok.

Other systems that followed hardly differ from the Seacon based systems in that most of the semi-industrialised building technologies are still limited to the skeleton system.

There are also companies that produce prefabricated floors. This started in 1968 when prefabricated clay pots were introduced. The technology was imported from Italy. The hollow clay pot acted as a void filler to reduce the dead load of the floor, but it had to be laid on wooden or steel formwork. The next development was the inverted T-beam, together with the concrete block floor. Many other types of prefabricated floor were also developed, e.g. hollow-core, U-type, inverted U-type and

Double-T, etc. There are currently more than 20 types of prefabricated floor.

The National Housing Authority of Thailand (NHA) has set up design guidelines based on modular co-ordination meant for both building design and site planning. At the same time the CI/SfB system was also introduced.

In 1976, three years after the NHA was established, the government set up a target to build 20,000 housing units annually for its low-income population. NHA had planned to construct as many standard apartment units as possible. At the same time research had been conducted to explore the possibility of pre-fabricated housing construction for approximately 10,000 one-room apartment units every year. The Seacon system was chosen for a medium rise apartment housing project at Din Daeng, Bangkok.

Following completion of the Seacon Apartment project, another low-rise housing project was immediately planned. This project, called the Tung Song Hong Project, is a modification of the Seacon System but with fewer, larger, components.

NHA did make an effort to experiment with semi-industrialised building systems. In 1984 a special committee was set up to explore the possibilities of utilising semi-industrialised building systems in the housing projects. This committee suggested that in future housing projects, if the contractors desire and have the capability to carry out the construction works in an industrialised system, they should be allowed to do so.

This suggestion was put into action in the Samutprakarn Project, another Sites & Services scheme. In this project NHA prepared the construction drawings for a conventional construction system, but it was added in the specifications that if desired, the contractors are allowed to propose and submit their own industrialised construction system for approval. This project was divided into several construction zones. Public tenders were called and bids were opened in April 1987. One of the lowest bidders proposed a components system for the construction of housing units. The prefabricated components were reinforced concrete columns, beams, walls and floors.

The private sector also moved slowly towards the industrialised building systems. Prefabricated floors have replaced the in-situ floors in most buildings which have repetitive floor spans. One of the reasons that make the prefabricated floors so popular is that most producers of reinforced concrete piles have added the production of prefabricated floors into their production plants. Since the production process is very similar, the reinforced concrete pile producers needed very little initial investment to produce prefabricated floor units. This has made the prefabricated floor inexpensive compared to the in-situ floor. The floor is the only prefabricated component that is widely used. The rest of the building components are constructed in conventional ways.

There are a few developers who have applied industrialised building systems in their projects. The half tunnel form system is an example. These projects were meant to publicise the system, but the system could hardly be feasible due to the high cost of the large formwork, not to mention the 40% import tax.

The development of industrialised building systems is slow in the private sector due to the country's economic and social conditions that still offer resources to be utilised more suitably by the conventional construction system. Still the concept for future development of industrialised building systems in the private sector is that the system has to be pulled by demand, not pushed by supply.

PROBLEMS RELATED TO SEMI-INDUSTRIALISED BUILDING SYSTEMS IN THAILAND:

1. In a country where labour cost is inexpensive, it is difficult to convince people, even those involved in the construction industry, to venture into industrialised building systems. It is a general perception that conventional construction systems can still be operated efficiently at present and in the near future. Some even ignore the possibilities of utilising the industrialised building system completely.
2. In a country that has very little experience in industrialised building systems, except for some staff in private companies who specialise in semi-industrialised building systems, as far as the industrialised building system is concerned almost all other construction people are classified as inexperienced project managers, inexperienced designers, inexperienced labourers, etc.
3. Investment in industrialised building system projects is considered a risk.
4. Industrialised building systems cannot be adopted from countries which have very different technical and cultural backgrounds.
5. In a country with a problem of unemployment, industrialised building systems should not mean just the replacement of man by machinery. It should mean a system made to utilise the existing resources more efficiently. Such a system takes great effort in terms of time and resources to experiment with it.

6. Local construction materials do not always conform to the same basic module. This is one of the causes of wastage that should be avoided in industrialised building systems.

Indonesia [27] (Partadinata)

The traditional semi-industrialised building system has long been familiar in Indonesia. Bamboo mats, door and window frames, foundation blocks, roof structures and roof tiles have been produced locally and traditionally.

During the last decade, Indonesia has implemented various semi-industrialised building systems as part of its national housing programme, such as particle board and wooden components, the BRE-precast concrete system, the Cortina-concrete system, the Alcan-Eternit sandwich system, etc. Since the standardisation of building material components or elements has not been fully adopted in Indonesia, standardised building components are still high cost products and are not yet accepted by the Indonesian people and its culture. Hence, the industrialised building system still has a long way to go.

The first use of pre-fabricated elements was concrete wall components for low cost housing projects the first quarter of this century. In the late 1950s the introduction and promotion of modular co-ordination and a modern industrialised building system had been launched through various seminars, workshops and lectures. In the 1970s, the rapid housing construction programme gave the opportunity to the industrialised or semi-industrialised building system to be practised.

Industrialised building systems have already been introduced in Indonesia, but few are in use. Some examples can be mentioned, such as mass production of houses using prefabricated particle board panels. Particle board panels are produced by two interconnected plants, the particle board plant and the wall panel workshop.

Standardisation has not yet been fully implemented in Indonesia. Various dimensions of wood are still found on the market.

Modular co-ordination is still a dream for an architect or researcher. There is still many difficulties in infrastructure and transportation, especially in remote areas and on the other islands of Java.

Malaysia [27] (Swee)

At present there are many industrialised building systems in Malaysia. In fact more than 140 system builders are registered with the government, most of which originated from the west. There are systems for small low-cost housing using light weight panels and systems for high rise flats using heavy panels lifted by tower cranes. Each category of system has its own peculiarity and use. Generally all the industrialised building systems in Malaysia can be classified under the following categories:

1. Large panel systems

Malaysia's first pilot industrialised building system using the Danish Larsen-Neilson system was undertaken in 1966 in Kuala Lumpur. A year later a second project was undertaken in Penang using the French Estiot System. The projects utilised the large panel system which requires large concrete panels cast in the factory and transported to the site on trailers for assembly. Erection was performed with the help of tower cranes. In the factory, two types of casting moulds were used: vertical and horizontal, made from steel. The speed of construction was much faster than the conventional method although the tendered price was slightly higher than the conventional method by about 5% to 8%.

Some recent projects completed using the large panel system in the state of Selagnor were undertaken by a joint venture company. More than 10,000 units were completed using this system. A circular factory was set up in Shah Alam for casting of the panels. Other large panel systems in Malaysia include:

1. Taisei Marubeni system - This is a Japanese system. Large panels are cast in a factory using a tilt-up system where one panel forms the base for the next panel cast. This system saves on plant cost. Some 4,880 units were built in Shah Alam using this system.
2. Precast system - One thousand and fifty (1,050) units of five storey low cost flats were built in Taman Brown. In this system pre-fabricated panels were used for floor and slabs; the panels themselves were used as moulds.
3. Ingeback System - This Swedish system used large panels for building construction. All the panels were cast in a central factory using vertical battery moulds and tilt-up table moulds.

The large panel system is normally confined to high-rise buildings and a large concentration of units close to the factory site. This system is the most used of the other industrialised systems available. It requires a high degree of organisation and is the most advanced form of concrete construction as it employs minimum labour. A set-back of the system is its unsuitability to the wet tropical climate of the country.

2. Metal form system

The system is more versatile and has been used on large and small building projects. There are no leakage problems and the initial capital investment is lower. Its additional advantage is the ability to adjust to suit architectural requirements.

Most of the builders using metal forms use the tunnel form system. This system uses both the half tunnel and a complete tunnel form; the half tunnel is composed of vertical and horizontal panels set at right angles and supported by struts and props. Positioning lugs and linking clamps ensure accurate connection of the formwork surfaces. Kicker forms are fixed on top of the tunnel units before casting of the concrete to obtain an exact position of the walls on the next level. The walls and slabs are cast in a single operation.

Three large projects in Wangsa Maju, Pandan Jaya and Taman Maluri used this system. The system leaves no joints and therefore leakages are unlikely to happen. The finished product is smooth and with hardly any honeycomb. The system cuts down on service expenses as fittings and services ducts are all cast in and small projects can easily be constructed.

Wall and table forms have been used locally in a number of projects. The wall forms are combined with the slab form so the walls and slabs could be formed monolithically in one casting operation. Although the system originated overseas, local contractors have made modifications to suit local requirements. Instead of steel, high quality film-coated plywood shuttering is used. The form can be easily dismantled and handled by small cranes and can be adjusted to suit architectural requirements.

3. Framing system

The system has been used only in three projects because project cost is higher than similar projects using the conventional system.

This system works on the same principle as conventional beam and column in-situ concrete systems. Here however, the building components are precast and transported to the site by trucks. This system lays claim to better efficiency in construction and a shorter period in construction time.

The columns are cast in steel moulds with lengths of three storey height, but before casting, RHS connectors are fixed to the mould for single casting, bolts and plate connectors are inserted in the mould for single casting, and bolts and plate connectors are inserted in the mould where column-to-column connection is required. Other building components cast at the casting yard include beams and floor slabs. For initial connection of the foundation, a pocket is cast in the foundation to accept the lower end of the column. Erection is done in sequence with the help of cranes. After the erection and connection of the building components the joints are grouted to give a monolithic finish. All connections between beams and floors are provided with ties which are welded on site to enable composite action between floors and beams. All walls are built with infill material. Because of the low low-rise structure, all lateral loads are carried by the column. The only set-back in this system compared with the conventional system is higher construction cost.

4. Partially precast system

The Australian-originated system is predominently used in low rise building. The Cemlock system was used by the National Housing Department in Pekan Selama.

The system is labour-intensive and makes use of cemboard sheeting on prefabricated framing.

The structure wall consists of three basic components:

1. Reinforced concrete columns constructed integrally with the end and party walls of the unit with a core filling of square hollow concrete blocks for speed.
2. Cement block masonry infill panels bonded to the reinforced concrete columns as the gable wall.
3. Prefabricated cement board clad load bearing wall panels.

The three components are arranged and connected together to the concrete floor slab to form a stable structure.

Another system used mainly for single storey buildings is the 'Q' built system. The essential feature in this system is the light-weight aluminium formwork. After the foundation has been constructed, mesh reinforcement for the wall is installed at the same time as the service pipes are put in place. The form panels are then secured in position by pins, wedges and clamps. The window frames, door joints and other special openings are fitted into the framework, after which concrete is poured.

The system has been used in a large number of projects because it combines conventional and system designs. The use of partial prefabricated systems will increase in the future with the establishment of hollow concrete block and prestressed joist industries.

5. Modular systems

The system reduces construction time and wastage of building materials. A high quality of finish is also obtained.

Complete or partially complete modules of dwelling units are prefabricated in the factory, and transported to the site and stacked and arranged into dwelling units. The system reduces construction time and wastage of building materials. A high quality of finish is also obtained.

6. Hollow core slab-reinforced concrete frame system

In recent years the production of hollow core slabs has started in large scale production plants. This means an open delivery principle because companies are supplying either components, like slabs, columns, beams or facade and partition wall units, or they are supplying structural assemblies like bearing frame-slab assemblies, partition wall assemblies or facade assemblies. Quite often the bearing frame is cast in situ and combined with precast prestressed hollow core slabs. No agreed dimensional co-ordination system exists, but the practice is to apply 1.2 m modulation, or less when needed.

Pakistan [27] (Maher)

The demand for low-cost housing is increasing. Conventional construction methods are not able to meet the demand due to the slow pace of construction and higher costs. Labour is still relatively cheap; however, the elimination of the middle man, through industrialisation of building components and direct use by consumers on a self-help basis, may help to reduce cost of building further. Partial prefabrication of building systems is therefore feasible in Pakistan.

Low income houses and some middle income houses are being constructed using the techniques of industrialisation to varying degrees, including ribbed slabs, precast tiles and battens (beams), precast and prestressed single tee and I-beams, hollow planks, precast panels of slabs, walls, sunshades, etc. The construction of industrial buildings utilises mostly precast beams and girders due to large spans.

Partially industrialised housing is gaining popularity in Pakistan due to savings in time and money. The housing uses precast beams and factory made tiles for roofing. Panellised housing has recently been introduced on an experimental basis. The houses are being constructed using precast panels for walls, slabs, stairs and sunshades. The connections consist of tongue and groove (for wall to wall), welding of protruding rods (for wall-slab, slab-slab, wall-plinth) and grouting.

Precast components including pretensioned single T, double T channels, I-beams, hollow planks, and precast tiles are factory produced in various sizes and spans to suit the requirements of individual buyers. The flush joints are grouted for single and double T beams, channels and hollow planks.

Philippines [27] (Commandante), (Aureus-Eugenio)

Among the more common types of industrialised building components used in present civil and structural projects are the prefabricated hollow core slabs, waffle slab system, double tees as floor and roof units and steel frame trusses. Prestressed pre-cast concrete is now the byword among architects and engineers. The prestressed and post-tensioned systems introduced in the Philippines as early as the 1960s is limited to floor and roof units. Structural steel has been more widely used, not only in commercial and industrial buildings, but in structural trusses for bridges and antenna towers.

The industrialised building systems in the Philippines, whilst several were introduced during the last decade, have not fully been developed to the point that their use could be totally viewed as factors adaptable to mass housing in terms of the criteria required by the government. The more popular building systems that were used in some of the government's projects were the W-panel and Pontalift .

The basic materials from which housing systems are commonly constructed do not differ significantly from those used in conventional construction. Included in the list are wood - including lumber and plywood; composite materials (sandwich panels with cores of paper honeycomb and facing of plywood), lightweight concrete and prestressed concrete. The structure types are mostly the bearing panel systems.

Transplanted industrial systems could not be used directly in the Philippines without some modifications. Differences in structural design approaches, field practices, and materials did not fully arouse public interest and acceptance. Panel systems employ the use of materials, which include wood framing, reinforced concrete and sandwich panels.

The industrialised building systems are classified according to material categories, structure types, panel types, fabricated location and housing type.

The Modular Thin-Shell Concrete House uses as basic materials fine aggregates, steel mesh, cement and bamboo lath. The structure consists of wall and arched roof modular design. The form makes the structure strong and durable. A 10% to 20% reduction in construction cost of roofing is achieved.

The Panel-Lock Housing System uses pre-cast reinforced concrete walls and a choice of either a pre-cast concrete roof or conventional roof trusses which can be quickly erected. The basic component consists of aluminium panel frames. They serve as forms for the pouring of the panel lock panels and as columns to support the roof.

The Hi-Citra Housing System makes use of exterior and interior wall frames and structural components. The wall frames are anchored by protruding bars to connecting columns. The joints between wall frames, infill wall panels and roof panels are pressure-injected with non-shrinking cement grout. The structure is believed to be strong and well-braced and well-suited to Philippine environmental conditions including typhoons and earthquakes.

A Hi-cetra house is concrete and the roofing material consists of corrugated asbestos sheets. The extensive application of mass production methods save man-hour and material waste. The system has a mobile factory in the immediate vicinity of the job site. The co-ordination between the casting crew and the assembly crew enables the speedy erection of housing units.

The Shell-Type House was introduced in 1975 as a low-cost shell house designed to be within the reach of lower-income families. The company manufactures the roofing material and the asbestos-based flat sheets from locally available sources. The costs are significantly lower compared with units using traditional materials.

Nepal [1]

Proper governmental policies and appropriate guidelines for the use of building materials are crucial in facilitating the housing process and insuring appropriate standards. After land, building materials are perhaps the most important inputs in creating housing in the third world. Many technical deficiencies, high cost and difficulty in building housing can be attributed to improper selection of building materials. Housing in Nepal is still individually and uniquely built. The materials used are few, and construction is done manually. Compared to the housing construction in developed countries, where varieties of manufactured components are available to be assembled on site, construction in Kathmandu starts from the basics providing opportunities for creating specifically tailored details.

Use of indigenous materials, such as bricks, Kacho Int, tiles, and bamboo etc. can be easily promoted.

India

Characteristics of building in India [37]

(a) Building materials

Building materials generally used in urban India are plain or reinforced concrete and bricks for foundations; bricks in cement mortar and/or reinforced cement concrete for walls; reinforced concrete for upper floors/roofs; plain cement concrete with terrazzo finish or stone for ground floors; timber for door/window frames and shutters; bricks for partition walls; and cement plaster for internal and external finish. Stones or cement concrete blocks replace bricks wherever the latter are of inferior quality. Other materials like asbestos sheets, galvanised iron sheets, lime mortars, plasters and concretes, timber trusses, and reinforced brick etc. are getting less popular, especially in the urban context.

(b) Aspects and goals of building technology

Building technology in India does not seem to have any specifically articulated goal(s). However, it is expected to:

- Increase the rate of construction of houses and other buildings;
- Provide as much employment as possible to unskilled, semi-skilled and skilled male & female workers;
- Improve the quality of construction of skills of workers;
- Lower the cost of construction through use of better design-construction management systems as well as through financial control and management methods.

(c) In use and operation

Small size full or partial prefabrication technologies provide real scope for open industrialisation. Since neither the capital cost involved in various moulds, tools, plants, and equipment is high, nor many factory-type sheds etc. are needed, there is an almost total flexibility in the size and shapes of building components (of course limited by the inherent characteristics of the materials used). However, quite frequently, joints in buildings constructed with prefabricated components do not get sealed properly, giving way to water leakage, and this problem becomes quite serious as small-size components automatically lead to increases in the number of joints.

(d) In maintenance & repair

In general, maintenance of buildings is of quite low standard and repairs are of poor quality in India. Preventive maintenance is seldom practised. In view of this, it can be easily inferred that buildings constructed with pre-fabricated components also suffer the same fate.
(e) Recycling & reuse

Not many known cases seem to exist where prefabricated components have been recycled and/or reused.

Open industrialisation

In India prefabrication techniques for large-size panels/components have not shown any promise. During the last two-and-a-half decades, only a few dwelling units (in 4 to 6 storey blocks) have been constructed with these systems in Bombay, Delhi and Madras. Various reasons have been put forward for this tardy progress. Some of them are: (i) The building construction industry in India is highly fragmented. Both vertical and horizontal co-ordination amongst various partners is completely missing; (ii) Infrastructure facilities (like good roads, transport vehicles, cranes etc.) are, qualitatively and quantitatively, grossly inadequate; (iii) Managerial skills required for design office to factory to construction site co-ordination are in short supply; (iv) Such constructions are costlier than in-situ constructions; and (v) Interplay between architectural design and living pattern of people is poorly understood.

Research and development activity has led to the evolution of small-scale full and/or partial prefabrication of concrete, steel, and brick based building components. In the Indian context, small-scale partial prefabrication provides many significant advantages over other prefabrication techniques and systems. Some of them are: (i) They do not reduce employment; (ii) They gradually upgrade skills of workers through informal on-site training; (iii) They provide the right mix of men and machines in an Indian context and are not capital intensive; (iv) From the locational viewpoint, these technologies are quite flexible and mobile; and (v) Building components manufactured by them are cost-competitive.

Trends in the development of building technology

(a) In design

Two significant trends which are emerging in architectural design in India are: First is a shift away from the designing of individual houses (or buildings) to that of whole groups or complexes; and second is a wider acceptance of repetitive design, also the use of increasingly advanced design tools for the design of fully or partially prefabricated components, especially in reinforced concrete and reinforced brick.

(b) In products and materials

Earlier it was common to handle, lift and assemble all building components only manually thereby restricting each component's weight to less than 100 kg. But now there is a trend of developing and using inexpensive tools and machines for similar operations leading to a relaxation of the limit of each component's weight. Also, there is increasing demand for prefabricated building components with a high quality initial finish. Full and partial prefabrication is also making in-roads into components of reinforced brick, steel, and composites etc.

(c) In production organisation and technology

Large-scale production of small-sized fully or partially prefabricated components have been tried under various situations, especially for house construction programmes prepared as disaster (earthquakes, floods, cyclones) relief measures, but with varying success. Consequently, the trend of either on-site prefabrication of centralised production at a limited scale continues. As a matter of fact, most of the prefabrication techniques developed during the last decade do not provide any significant economies of scale.

Results of an inquiry for the state of art analysis of industrialised building (Misra):

Building technology used for construction varies widely throughout India depending upon geo-climatic conditions of each region, socio-cultural and economic situations, governmental policies and actions, rates of economic development and the distribution pattern of overall material gains of development. However, during the last decades, a perceptible shift is evident in the type of materials used in buildings: i.e. from mud, timber, bamboo and thatch towards burnt brick, cement, steel and stone, etc. The techniques of construction and manpower skills are also getting gradually upgraded.

Apartment buildings:

Open building design systems (dimensional co-ordination, tolerances, compatible components and joints) are used in the bearing frame 0% - 20%, the share is increasing; facade 80% - 100% decreasing; roofing 80% - 100% stable; partitions 20% - 40% increasing, and building services (installations) 0% - 20% stable. Full prefabrication of product systems, modules and components is not applicable, but among composites (prefab and in situ) the bearing frame 0% - 20% decreasing; facade 20% - 40% stable; roofing 80% - 100% increasing, partitions 20% - 40% increasing and building services 0 - 20% stable. Open computer aided design systems are not applicable. Modern methods are used in all construction parts 20% - 40% stable, but in building services 0% - 20 % and stable.

Office buildings:

Open building design systems are used in bearing frames 60% - 80%, the share is increasing, facade 0% - 20% stable, roofing 80% - 100% stable, partitions 20% - 40% increasing and building services 0% - 20% stable. Full prefabrication of product systems, modules and components is not applicable. Composites are used in the bearing frame 40% - 60% increasing, facade 0% - 20% stable, roofing 80% - 100% increasing, partitions 20% - 40% increasing, building services 0% - 20% stable. Open computer aided design systems are not applicable. Modern methods of production planning and control are used in the bearing frame 20% - 40% increasing, facade and roofing 20% - 40% stable, partitions and building services 0% - 20% stable. Systematic quality assurance systems (ISO 9000) and advanced modular production infrastructure and organisation (construction with effective use of external deliveries and services) are not applicable.

One family houses and detached buildings (row houses):

Open building design systems are used in the bearing frame 0% - 20% stable, facade and roofing 80% - 100% increasing, partitions and building services 0% - 20 % stable. Full prefabrication is not applicable and composites are used in the bearing frame, partitions and building services 0% - 20 % stable, facade and roofing 80% - 100% increasing. Modern methods are used in the bearing frame 20% - 40% stable, facade and roofing 40% - 60% increasing and partitions and building services 0% - 20% stable. Systematic quality assurance systems and advanced modular production infrastructure and organisation are not applicable.

Traditionally, rural housing in developing countries is based on indigenous building materials and construction techniques which are in harmony with nature. In a rapidly changing context, it is becoming increasingly difficult for the rural masses to retain traditional building methods. There is a change in the perception of building construction in rural societies.

A study of building materials selected without considering the affordability factor indicates that 95.2% preferred burnt brick walls. As roofing materials, 80.6% preferred a combination of dressed wood beams, locally available 'Cuddapah' stone slabs and mud, which is an improved version over traditional mud roof on rich landlords' houses in the region. To avoid seepage, they prefer to lay stones as a damproof course. Since the roof receives maximum exposure from the sun during the day, they prefer to retain a thick layer of mud over the stone slabs to act as insulation. Respondents were not in favour of reinforced concrete slab. They are aware that concrete is a bad heat insulator and it would require artificial cooling or ventilation to keep a concrete house comfortable. The majority of peasants live in mud houses because they have no other choice. They cannot afford to pay for the cost of burnt brick walls and dressed wood roof [44].

Use of unburnt soil bricks for building walls is a traditional practice in the region, therefore stabilising the soil and compressing it in a manually operated press could be considered an affordable improvement over the existing techniques used. When respondents were presented with information using flash cards on stabilised and compressed soil bricks, 85.8% expressed their willingness to adopt this new technology provided it was within their means. Similarly, when they were explained that pre-cast reinforced concrete joists would be a cost effective replacement for wooden joists, they indicated their willingness to use them if they were made available [44].

Research has been done concerning the shapes of industrialisation of urban housing construction in India, with the big cities of Delhi, Bombay and Madras as study areas. By industrialisation we mean mainly use of precast concrete elements and mechanisation of the site. It seems that the most suitable shape is an industrialisation made of small and commonplace products and components. The best examples of these products are concrete hollow blocks or lintels. It appears that industrialisation is a means to increase the present level of productivity and quality [24].

Taiwan

It was not until the late 1960s that building industrialisation was initiated in Taiwan. Obviously, satisfying housing needs was the main focus for housing policy, due to a combination of a recovered economy and the already substantially expanded population. And this is clearly the objective of building industrialisation during this time. The first attempt was earmarked by the government office responsible for economic development, with co-operation from a major constructor in Taiwan also owned by the government. In 1971, the first Sino-Japanese prefabrication plant was formed. From 1973, a significant series of research reports documented a variety of prefabrication-related concepts and technologies. In the following several years, at least a total of 12 firms invested in either the prefabrication plant or other prefabrication-related technologies. Not surprisingly, the application of prefabrication technology was primarily towards the construction of government funded public housing. By 1980, a huge number of industrialised buildings were completed, most of which were 4 - 5 storeys high. Only a few rose as high as 12 storeys.

The burgeoning of the prefabrication industry was apparently headed by government policy at that time. Of course, the building investment was also chiefly injected by government funds, perhaps some of which were from the U.S. Close to the late 1970s, most industrialised buildings were completed. After a short period of adaptation, most house owners were unsatisfied with the prefabricated product. The main complaints documented include the following:

1. Leakage in joints and roof
2. Noise due to improper joint seals
3. Lack of flexibility to change interior partitions
4. Inefficient interior design
5. A sense of low quality due to inconsistent tolerances

The market of unsold or construction-in-progress prefabricated building units was stagnant. Impacted by the lack of cash inflow and the diminished future market, literally all building prefabrication firms went bankrupt. Those that barely survived did so either because of government ownership or thanks to compensation from other more lucrative businesses. Although the development of building industrialisation is still in its infancy in Taiwan, this failure quickly buried what was learned, and R&D related to prefabrication thereafter seems only restricted to academic environments.

In the second half of the 1980s Taiwan's economy reached its highest peak. The construction industry responded with a sudden outburst of a great number of new firms. The rapid transition starting from the second half of the 1980s may seem optimistic for all construction practitioners, including investors, designers, general contractors, subcontractors, material suppliers and construction workers at first glance. The government was also satisfied with this development for a brief period.

Facing these difficulties, the National Construction Automation Plan was set forth by the government during the turn of 1990 to 1991. Many R&D collaborations were executed at full speed. However, building industrialisation was not on the priority list in the early stage of the NCAP.

The main approach of building industrialisation is to construct a building from prefabricated items. When this concept is translated into construction of the main body of the building, the constructor would have to make decisions of how the following items are produced: slabs, columns, beams, exterior walls, interior walls, stairs and balconies. In the same manner, the following items should be considered for the non-body portion: exterior finishing, interior finishing, partition walls, door/window systems, toilets, kitchen and other functional facilities, elevators, instrumentation and other functional equipment.

At the current stage in Taiwan, the following philosophical debates still prevail among major developers:

1. Can prefabrication shorten the construction schedule and therefore speed up sales and revenue?
2. How would prefabrication influence building marketability?
3. Can prefabrication meet users' functional requirements?
4. Can prefabrication reduce construction costs?

The newly employed methodology is basically rationalisation or fusion of site-based and prefabrication based methods. Compared with the approach taken in the 1970s, several conceptual strategies employed by this series of efforts are worth mentioning.

1. Mechanisation
2. Mass Production
3. Standardisation

The concept of standardisation covers three specific areas. The standardisation of dimensions in design and building layout within the same project enables the reuse of forms and supporting scaffolds, reduces possible imprecision or construction errors, and increases the quantity of identical building items in a project. The standardisation of construction methods and erection/connection/assembly procedures shortens the learning curve, economises the use of learned know-how and minimises the cost of communication among parties. Standardisation of the use of building material as well as the associated fixtures simplifies the whole design and the related construction technology. The product of realising these strategies is a hybrid building construction approach, which combines both the concepts of productivity improvement and building industrialisation. Specifically, method improvement tools and prefabrication methods are employed in parallel to meet or satisfy the following objectives or constraints:

1. Environment/Site Conditions
2. Economic Conditions
3. Technological Requirements
4. Societal Conditions
5. Legal Aspect
6. Design-construct Integration

The rationalisation of construction method innovation can be envisaged from two aspects. The first examines the appropriateness of how high-value-added building items are constructed.

The other pertains to the reduction of labour reliance. This can be achieved by reckoning the relative percentage of the type of labour usage in similar projects which were constructed by the conventional site-based method. Form work, plastering and repair work are identified as the top three activity items which rely on the employment of labour. Benchmarks can be assigned to each category for labour reduction.

After deciding the construction road map and the rationalisation of the hybrid construction approach, the design-construct team has to conduct a thorough survey of the status of industrialised building items available on the market. At times, the team needs to issue inquiry to the manufacturers concerning the production and material delivery schedule, and quality differentiation of different production batches. If building items are to be prefabricated at an off-site location, the team has to study the likely production schedule with respect to the off-site area available, weather conditions, transporting conditions, etc.

A final step relates to drafting the construction plan. From the limited number of projects examined, there are three main themes to be dealt with in the construction plan. The design-construct team has to identify the activities on the critical path and begin to level the manpower required throughout the project schedule.

A second theme relates to the lifting plan of all prefabricated items as well as the associated large/heavy materials. To draft the plan, the team needs to follow the master construction schedule and the bills of quantity of all large/heavy prefabricated items.

A third theme covers the interrelationship between the detailed prefabrication design and the production schedule. The division of prefabricated items should hide all seals and allow maximum efficiency in production and lifting.

Building industrialisation is one major area of construction automation for Taiwan in the next few years. Now wiser from the failures of the 1970s, a considerable number of constructors are exerting serious effort in trying to economise the fusion of the conventional site-based construction method and the prefabrication-based one. Preliminary successes have been documented and a few more can be foreseen. However, it seems to the writer that the importation of foreign labour and a mild recession from 1994 have slowed down the evolution of the hybrid construction approach described in the paper. Nevertheless, the trend of innovation will continue until another level of technology maturity is accomplished.

3.2.5 Northern America

Canada [34]

Market demand is undoubtedly the most powerful force for change in the development industry, and it is clear that if there were a strong demand for green buildings, developers would change the face of our cities in short order. One reason why this is not occurring is the current low level of market demand for any new construction; a total lack of demand for office projects and low demand for housing. This situation relates mainly to economic and demographic conditions, and it can be assumed that this situation will gradually improve over the next few years.

It is widely recognised that existing buildings are relatively inefficient in their energy use. New buildings tend to reach higher levels of performance, but usually within the same order of magnitude as existing buildings. Since most buildings are directly or indirectly dependent on the use of fossil fuels, the rate of depletion of these resourses is a concern, as are the collateral negative impacts on the natural environment. In neither the office nor residential markets is there a clear demand for other important facts of building performance: impact on urban developments, energy performance, ecological performance, other aspects of indoor environment performance, such as lighting or acoustic quality, adaptability to change, or the ability to maintain performance over the long term. This is not surprising, since neither office (except for very large organisations) nor residential tenants can be expected to make themselves expert in the nuances of building performance. The Building Performance Assessment System, or BEPAC, developed by Ray Cole and others at the University of British Columbia, is an example of a system that has the potential of providing commercial tenants with the tools needed to differentiate between buildings of different performance levels; and if such a system becomes widely adopted the influence on the market could be very significant.

Governments have attempted to stimulate demand by sponsoring demonstrations of new ways of building, but it must be recognised that these may have limited impact on the general market. Demonstrations of high-performance houses is a well-proven method of stimulating consumer demand for new housing features, and exhibition display houses, or more serious efforts such as CANMET's Advanced Houses programme, undoubtedly have some impact on the housing market.

Contemporary demonstration programmes in the large-buildings sector, such as CANMET's C-2000 programme or the CMHC/CANMET Ideas Challenge programme, show that high-quality and desirable buildings can be designed and built with advanced environmental features and with energy consumption levels of about 50% of current good practice, all for very modest levels of incremental costs. Demonstrations of commercial buildings, however, are more effective in influencing developers and design professionals than the real end-users. An attempt to influence the large-building sector is therefore likely to require a combination of demonstrations and efforts to provide commercial tenants with the right tools for selecting green buildings.

Possible approaches

Substantive retrofit measures can and should be applied on a wide scale, but even the successful and widespread adoption of retrofit programmes will not solve the dependence on the automobile caused by our current low-density development pattern. In any case, retrofitting is not likely to suffice, since a significant proportion of the existing building stock is economically or technically suited to being upgraded to only modest levels of performance. A significant amount of new construction may therefore be needed to meet the needs of 2025, much of it replacement for existing low-density construction that cannot be upgraded to required levels.

In either case, whether new or retrofit, buildings will have to reach very high levels of energy and environmental performance, while stung and density decisions will have to be aimed at reducing the dependence on the automobile. Given the nature of the problem, what practical and widespread measures are available to achieve this?

Building performance standards

The financial and regulatory approaches described above all make reference to various levels of building performance. The successful implementation of these concepts will require a comprehensive and tested reference standard of building performance, including a definition of parameters to be included and levels of performance to be attained in each area. Although the main concern in this paper is energy and environmental performance, it must be recognised that private-sector developers and designers have broader concerns, including air quality, case of maintenance, cost-effectiveness, etc. The probability of market acceptance is therefore increased if the performance standard covers a wide range of issues. The CANMET C-2000 programme has successfully demonstrated this approach. The C-2000 criteria may serve the required purpose of defining required performance levels, and the six demonstration projects being carried out under this programme can serve as performance benchmarks.

The specific performance areas applicable to building design (and excluding, for the moment, location issues) that would be covered in such a broad and market-oriented framework include the following:

- energy performance, specifically maximum annual energy consumption and demand,
- ecological performance, including impact on local ecosystems, building emissions, resource conservation, and waste minimisation,
- indoor health and comfort, including air, lighting and acoustic quality,
- appropriateness of design, spaces and systems to functional requirements,
- longevity of the building and systems,
- adaptability of the design, spaces and systems to changing requirements,
- durability of components and materials,
- case of maintenance and operations,
- quality management during design, construction and operation, and
- economic viability.

Most of the performance parameters listed above are obvious, but the inclusion of longevity, adaptability and durability deserves an explanation. All other factors being equal, long-lived of a building clearly reduce the environmental impact due to the production of new materials and the disposal of old ones, in addition to the energy cost of the construction process itself. Components and materials must be durable to support long life, but this will only be successful if the building and as major elements are adaptable to change over time. Adaptability is thus a critical requirement for good environmental performance, and extends to changing uses in the building as well as alterations required over time in equipment or materials.

Performance-linked incentives can only be effective if they are implemented within a framework that will ensure that requirements are actually reached. Application on a broad scale would require the adjudication of initial and on-going performance with reference to the performance standard by an independent third party, such as a governmental or non-profit organisation with no direct economic interest in the results. In such a scenario, projects that did not reach the required performance level would not gain the certification needed to benefit from the tax benefits proposed. Again, the C-2000 programme has shown that an approach that establishes a challenge for design teams can yield very successful results, without destroying the efficacy of the normal development process.

Open building systems

The discussion above suggests that policies to support long-term high performance in buildings should give encouragement to large buildings and large organisations; but we need not limit ourselves to traditional interpretations of scale. One way of reconciling the need for large structures with the desire for smaller, more adaptable and more humane buildings is to follow the lead developed for residential buildings by N. Habraken and the Open Building Foundation in the Netherlands. The concept is known as Open Building Systems and the idea can be extended to other building types. In such an approach a separation can be made between large-scale and long-lived base-building structures and the more varied and shorter-lived systems and facilities contained within it. The large and permanent structure is custom-designed for the site, while shorter-lived internal systems are manufactured off-site to modular dimensions and installed as a quite separate operation.

An example worth mentioning is a highly advanced craneless method for building precast houses. The proposed method for building single and two-storey precast houses is based on patented mobile machinery equipped with pivoting and lifting forms which produce and erect wall panels and floor slabs [25]. This approach offers another distinct advantage, in that it offers the possibility of greatly extending the life-span of buildings, by facilitating functional changes within the base building without requiring major renovations. This has a clearly beneficial effect in the reduction of future need for new construction materials, and in the disposal of construction and demolition wastes. Such an approach also enhances the long-term value of the asset for the owner.

To address the issues discussed above, a logical approach would be to develop a legal and tax framework that would encourage the private sector to design, construct and operate very high-performance buildings. Such a framework will have to be designed around the points already raised, namely:

- the preparation of enabling federal and provincial legislation to create a class of development corporations that can benefit from investor tax incentives and accelerated asset depreciation measures,
- the development of a two-tiered and performance-based property taxation structure and the adoption of more mixed-use, medium-density zoning techniques by municipalities,
- the application of comprehensive and well-tested building performance criteria,

- the designation of an independent third party to adjudicate the attainment of performance,
- limitation of eligibility to development and building management organisations with proven professional skills in building design and operation,
- limitation to buildings that are designed for mixed uses and medium densities,
- limitation to buildings that are designed for mixed uses and medium densities,
- limitation to buildings that are large enough to ensure that alternative design analysis and performance simulations can be economically carried out, and
- the simultaneous implementation of a widespread system of building performance labelling, so that tenants will be induced to select high-performance projects.

The new class of organisation that would emerge from this framework, referred to here as a Green Development Corporation (GDC), would have a well-characterised structure and predictable mode of operation, with an emphasis on long-term profitability and the development of energy-efficient and environmentally sustainable development projects. Our goal is that such corporations would become widely established by private sector developers.

Conclusions

Although the current lack of construction activity may lead to the conclusion that immediate action is not needed, the length of time required to achieve meaningful change in our urban fabric means that there is only a short respite. We can no longer afford the luxury of designing better single-detached forms of suburban housing, or communes designed for rural settings. Such approaches may be inspirational, but do not address the issues of conservation of agricultural land or the inefficiencies of low-density development. If the challenges of the twenty-first century are to be met, it will have to be through urban solutions designed to make our major cities better and more efficient places sites for work, commerce, housing and play.

USA

The U.S. construction industry is the nation's largest single industry. In spite of the economic significance of the industry and in spite of the high societal value of a home, the housing industry has stayed technologically stagnant in the USA. The residential construction industry in the United States has a short range planning tradition and is very fragmented. This "small business" structure is not conducive to industry funded technological research. A major constraint in the technological evolution of the U.S. housing industry has been the lack of a centralised R&D effort. The building industry has attracted very little or no public funding for research. One of the several reasons for the slow evolution of new technologies in the residential construction is the difficulty in bringing new materials and methods through the maze of codes and regulations. Separate building codes for each state, country or city have no rationale [38].

Current home building practices in the U.S. vary widely; there can be as many practices in a region as there are builders. A well researched best construction practice does not exist because there is no centralised research co-ordinating agency and because the industry is highly fragmented. A wide variety of materials are used in residential construction. Precast concrete has not been utilised to any great extent in the housing industry [38].

New technologies to improve the productivity and the product being delivered to the builders need to be investigated. These include such areas as robotics, panelised construction, and efficient fastening and connection techniques. Robotics can be used in hazardous operations like deep-trench excavation and tunnel boring, and in repetitive work such as roofing and brick laying [38].

The most vexing problem with low-income precast housing has been the rigidity of the forms; when tailored specifically for one design, it will practically exclude all other options. This rigidity forces a developer to amortise a large capital investment on a single project. Most projects cannot carry the burden of such large expenditures. Proof of this dilemma is the very limited use of pre-casting in the housing industry (currently estimated at less than 2% in the USA). The use of five-sided modules, such as the IFC system, provides a modest increase in flexibility of design, along with a decrease in time [40].

Industrialisation of single family housing in the United States [11]

Housing represented 55% of the total expenditure for building construction in the U.S. in 1994. Of the $179 billion housing constructed in 1994, 10.5% of the expenditures were multifamily, 3.5% HUD code and 86% single family (U.S. Bureau of Census, 1994).

Of the many definitions currently used to describe industrialised housing, the four used here are:

(1) **HUD code** houses (mobile homes)
(2) **modular** houses
(3) **panellised** houses (including domes, pre-cut timber pieces, and log houses)
(4) **production-built** houses (including those that use only a few industrialised parts).

These four definitions were selected because they are the categories used to collect statistical data, and so are likely to persist. However, the categories are confusing because they are based on a mix of characteristics: unit of construction (modular, panellised), method of construction (production-built), material (panellised), and governing code (HUD Code).

A HUD code house is a movable or mobile dwelling constructed for year-round living, manufactured to the pre-emptive Manufactured Housing Construction and Safety Standard of 1974. Each unit is manufactured and towed on its own chassis, then connected to a foundation and utilities on site. A HUD code house can consist of one, two, or more units, each of which is shipped separately but designed to be joined as one unit at the site. Individual units and parts of units may be folded, collapsed or telescoped during shipment to the site.

Modular housing is built from self-supporting, three-dimensional house sections intended to be assembled as whole houses. Modules may be stacked to make multistorey structures and/or attached in rows. Modular houses are permanently attached to foundations and comply with local building codes.

Panellised houses are whole houses built from manufactured roof, floor and wall panels designed for assembly after delivery to a site. Within this category are several sub-categories. *Framed panels* are typically stick-framed, carrying structural loads through a frame as well as the sheathing. **Open-framed panels** are sheathed on the exterior only and completed on site with interior finishes and electrical and mechanical systems. *Closed-framed* panels are sheathed on both the exterior and interior and are often pre-wired, insulated and plumbed. *Stressed-skin panels* are often foam filled, carrying structural loads in the sheathing layers of the panel only.

Production building refers to the mass production of whole houses "in situ". This large and influential industry segment is industrialised in the sense that it employs rationalised and integrated management, scheduling, and production processes, as well as factory-made components. In this instance, however, rather than the house being built in the factory and moved to the site the factory is the building site, which becomes an open-air assembly line through which industrialised labour and materials move.

These categories include a wide variation of building methods and building types.

In the past 10 years, panellised construction has increased its market share from 30% to 39%, HUD code housing has increased from 16% to 19%, modular buildings have been relatively stable at about 6%, and production buildings have declined. In numbers of units, panellised construction has increased from 540,000 to 679,000, HUD housing from 283,000 to 334,000, modular from 77,000 to 109,000, and production builders from 907,000 to 627,000.

U.S. housing production is becoming increasingly industrialised with the more industrialised HUD code, modular and panel producers taking the market share from the less industrialised production builders.

While housing is becoming more industrialised it is important to realise that the level of industrialisation within factories in the U.S. is low compared to other countries with very little capital expenditure on equipment and a reliance on inexpensive labour that can be laid off in response to declining sales which result from the cyclical nature of the U.S. housing market.

Almost all single family buildings are constructed from wood. Typical construction would include wood trusses for roofs, studs for walls, and joists for floors. Variations in construction occur regionally in the U.S. caused by culture, climate, and market differences. For example, in the case of floors in the south, 59.2% of houses are built on concrete slabs, 23.1% on crawl spaces, and 17.8% on basements. In the Midwest, however, only 10% of houses are built on slab, 11% on crawl spaces, and the majority (78%) are built on basements (U.S. Bureau of Census, 1994).

In the past decade, energy use in residential buildings has been a major consideration in some areas of the U.S. For example, HUD coded housing which has traditionally had very poor performance has seen energy savings of 60% in the Pacific Northwest as a result of the MAP and Super Good Cents programmes.

"A building system permitting the use of several alternative components and assemblies. The compatibility of components and assemblies is assured by means of a dimensional and tolerance system and joints. The system is open to free design and competition between suppliers."

Industrialised single family construction in the U.S. has certain characteristics of an open industrialisation system, but not others. U.S. construction is very open in terms of there being multiple components for housing manufacturers to select from, all of which fit into the system. For example, there is a tremendous range of windows available that vary widely in terms of cost, performance and material, all of which work in residential construction. The U.S. system, however, lacks a method of dimensional co-ordination on a small scale and instead depends on a composition of wood which can be cut and machined to achieve the connection and tolerance needed between other components. On a larger scale there are accepted dimensions, for example the 16" or 24" spacing of studs and the 4' x 8' dimensions of sheet products like plywood that allow the easy assembly of walls, roof and floor, but this is hardly a "snap together" system.

Some examples of industrialised building systems

A leading development in the U.S. prefabrication industry is an economical and flexible 3-dimensional prefabrication system, the TECHNOFORM machine from IFC. This hydraulically operated system casts five sides (four walls and a roof) in a single monolithic pour. A complete house shell is poured, cured and self-stripped within hours. Prototype low-income housing is currently being built in Puerto-Rico with shells giving 60 m^2 floor plans. The wall thickness of the shell is normally 75 mm, with a roof (which becomes a second floor slab) thickness of 100 mm. The walls are cast with windows and doors in place; plumbing and electrical ducting can also be cast in as an option [40].

Precast concrete in the USA is currently the most economical solution for warehouses, parking decks and garages. It has also gained economic importance for jails and prisons. For the latter, precast concrete produces a monolithic cell with uniform qualities and reduced construction costs, since the same cell dimensions can be used for administrative and support facilities. Precast cells have been stacked up to 14 storeys high although four storeys is the norm. The structure is quickly erected by stacking cells around common areas which are then roofed over [40].

The IFC machine has evolved from building small five-sided concrete structures, such as storm water catch basins, culverts and man-holes, and has grown to larger sewerage lift-stations, electrical transformer vaults, and double-celled prison modules. The latter have been as large as 4.5 m x 5 m x 3 m and included electrical, plumbing and security systems. The form machine itself, however, has been set up at the job site, close to the final placement site of the concrete modules [40].

The AIBS (ATLSS Integrated Building Systems) programme was developed to co-ordinate several research projects in automated construction and connecting systems. The objective of this programme is to design, fabricate, erect, and evaluate cost-effective building systems with a focus on providing a computer integrated approach to these activities. A family of structural systems, called ATLSS connections, is being developed with enhanced fabrication and erection characteristics. These ATLSS connections possess the capability of being erected by automated construction techniques. The technology for automated construction is heavily dependent on the use of Stewart platform cranes [50].

In the U.S. Shelley system the box units are arranged in a "checkerboard" fashion. The box units create additional space between them, which can also be used by closing it with wall panels, but requires a large amount of finishing work on site [52].

The Blakeslee method is a very simple version of an open/closed system. The system is basically adapted to rectangular layouts with a modular design grid of 3M x 12M. Its basic floor component is a hollow core prestressed slab 300 mm thick, 1.20 or 2.40 m wide, and with varying length of up to 12 m in 300 mm increments. The net floor height is 2.40 m. Interior load-bearing wall panels and exterior wall panels also conform to the basic design module of 1.20 m x 300 mm [52].

3.2.6 South and Central America

Argentina

The national housing plan, PNVU, required the construction of housing at a rate twice as large yearly as the best year on record, which was 80,600 units in 1955. Among the first projects funded in July in 1973 was a 350 unit housing project for army personnel which surprised everyone when finished within 5 months. They used 10 Outinord system forms. This encouraged a large number of contractors to associate themselves with national and foreign prefabrication specialists. By early 1975 there were five Argentinian systems in operation (Covipre, Polhuys, ERAS, Integral AS and Pinazo). In addition there were 16 European systems under use or testing (Allbeton, Malmo, Skarne and Elcon from Sweden; Vipresa from Spain; Camus, Coignet and Outinord from France; Peikert from Switzerland; Balancy from Italy; Bison from Great Britain; Larsen & Nielsen from Denmark; B. M. B. from the Netherlands; Okal-Haus, Elementbau and Estiot/Hochtief from Germany).

Most of these systems experienced economic difficulties. Prefabrication can only survive as an important ingredient in a national housing plan if given long-term projects (e.g.10 years) that permit ample amortisation of plant costs [41].

3.2.7 Africa

While the need for industrialisation in housing is unquestionable, the developing countries should be cautious in their choice of level of mechanisation, as many of them are already facing serious socio-economic problems with increased industrialisation in the housing and building industries. A number of case-studies have shown that, for both developed and developing countries, partial industrialised systems seem to offer a viable and satisfying solution in the promotion of mass housing. This applies particularly to considerations of long term planning suitable to future extension possibilities and maintenance needs of the community. Partial industrialised systems facilitate promotion of user participation in the development programmes, attracting adoption of self-built techniques which offer good economic flexibility [43].

Nigeria

Clay brick was a very popular building material in Nigeria during colonial times. In the present post-colonial era conscious efforts have been made to resuscitate interest in the use of bricks in building construction, but prospective house owners still prefer sandcrete blocks. Since the introduction of Portland cement sandcrete blocks have become the most popular material for domestic wall construction [28].

Since Nigeria gained independence, a host of advanced technologies has been introduced into the construction industry. Reinforced concrete, steel, lightweight plastic materials, sandwich panels, laminated boards and glass have become popular. These modern materials compete with clay bricks and sandcrete blocks in wall construction. Also coming into vogue are lightweight prefabricated buildings built with plywood boards, and synthetic boards are also becoming popular. Steel, reinforced concrete, and lightweight panellised reflective glass are currently used extensively in the construction of multi-storey buildings, offices and commercial buildings. As in North America and Europe, the tinted, reflective, heat-absorbent, sun-cooler "glass house" concept is fast becoming a fad. These glass and plastic materials are becoming increasingly preferred in the construction of curtain walls, panels and infill panels for large commercial buildings, because of their lightweight and space-saving attributes. Nevertheless, sandcrete blocks and clay bricks are still the materials of choice for domestic wall construction.

From early times clay brick has been a much used construction material. Traditional African society has always considered mud brick a close second to unprocessed timber as the best building material [28].

The construction process in Nigeria is basically traditional, characterised by on site and wet fabrication of components with gradual introduction of pre-fabricated components in the fields of joinery and carpentry, windows, doors and roof coverings. This has particularly aided faster construction in these trades. It will be of tremendous cost advantage if the units can be built en-masse rather than as one-off units. This further introduces the constraint that the units be built by a common agency if the cost advantage of mass production is to be gained [5].

Sandcrete construction is really in vogue in Nigeria, and a housing programme that uses alternative solutions is considered unacceptable. Cement being a major component part of any sandcrete construction makes this option a particularly expensive one [5].

Laterite, which is the most traditional and easily accessible building material in third world countries, has become a symbol of poverty and therefore not much encouraged by governments. Evidently, bricks and cement cannot satisfy Nigeria's housing needs. First there is no money for this, and secondly, there is a scarcity of such materials. Therefore the main criterion for the choice of material for housing low income people should be economy of construction [5].

Pre-fabricated trussed rafters in Nigeria

The major materials of trussed rafter are timber, plywood and connectors. At present in Nigeria there are some six firms that produce

78

plywood. The major raw material needed – timber – is abundant. It is recommended that the use of pre-fabricated trussed rafters be encouraged in the Nigerian building industry [6].

3.2.8 Middle East

Kuwait

The present building systems used in Kuwait's public housing projects suffer from long construction times and below-average quality. As a result, families experience long waiting periods and costly modifications and repairs. One way of decreasing project duration is using prefabricated building systems [8].

Only one housing project is known to have used the precast building system. The project started in 1976, when only one precast systems company was operating in the country. That experience was not successful as the duration of the project exceeded that of traditional building systems. Water leakage problems also occurred. Since then precast companies have not been involved in government housing projects [8].

The boundary wall system comprises a continuous strap footing, a retaining wall and columns. The width of the base and its reinforcement depends on the value of H which represents the difference between the natural ground level outside the house and the level inside the house. This system is somehow strange and rarely used [20].

Until 1985, buildings of housing projects in Kuwait consisted of reinforced concrete skeletons of beams and columns supporting slabs. Curtain walls of concrete covered with sand-lime bricks were then added. Since 1985 flat slab construction has been the method of choice. Here the beams are eliminated to allow for modifications often made by the recipients of these houses. Prefabricated building systems have been used in only one project [7].

In Kuwait there are six companies specialising in precast work. They act as main contractors or subcontractors or sell units. While they concentrate on the production and erection of precast units, in many cases they also use in cast-in-situ concrete construction. These companies produce a wide range of precast elements. Prestressing and post-tensioning of structural elements are also used in some of their projects. Other companies are either involved with cast-in-situ concrete or have just started a precast operation [7].

Libya

Libyans are familiar with one or two-storey houses and with a human scale of construction. The commonest feature of the new projects is their similarity in most Libyan cities and towns. Architecturally, houses have gained new design features. Courtyards, large windows facing the front and a vast range of imported decorative elements are considered desirable features of modern life [53].

However, and because the general policy of the country has focused on industrialisation and cultivation, large scale housing construction has been needed to encourage people to settle in agricultural and industrial projects. Up to 1988 14,000 housing units were constructed by the Agricultural Housing Institutes. These projects are dotted around the country. In addition, a large number of housing units have been erected beside new industrial projects such as the Suk Al-Khamis cement factory [53].

Saudi - Arabia

In Saudi Arabia industrialisation has been effectively applied since the 1970s. Huge building projects, often producing entire new cities, have been constructed. Several contractors from Europe have brought in their own technology, precast technologies and element factories. Contractors from the United States and Far East (e.g. Korea) have also been active, but more in the area of traditional cast-in-situ technologies. These factors combined have meant significant strides forward for the Saudi building technology. Concrete construction, either as large panel, filigran and hollow core slab prefabrication systems or as in-situ concreting, has become largely predominant, although steel structural systems also exist. Following the economic boom of the 1970s and 1980s, prefabrication and contractor companies have gained greater independence from their original owners. Although they use their own resources to carry out and further construction, they are doing so largely in co-operation with former owners or other foreign countries.

4

Conclusions and future trends

4.1 STAGES OF INDUSTRIALISATION IN DIFFERENT REGIONS

On analysis of the regional and national reports above, it is possible to classify the stages of industrialisation as follows:

1. Use of local materials in manual building on site.
2. Use of industrial materials in manual building works on site.
3. Use of industrial materials and equipment in manual building works on site.
4. Manufacture of building components partly in factories.
5. Application of closed building concepts with prefabricated components and modules.
6. Application of closed building concepts with prefabricated components and modules together with computer aided design and production.
7. Application of open building systems with alternative prefabricated components and modules from different suppliers together with computer aided design and production.
8. Application of open building concepts producing the deliveries (components, modules and systems, as well as development, design, assembly and finishing services) in networks of companies. The building concept includes the products as well as rules and guidelines for product systems for design, production and use.

Using this classification the regions can be allocated roughly as follows:

Region	Stage of	General trend of industrialisation in building
1. EU and Europe	4 - 8	Stepwise further industrialisation and automation in manufacture. Rapidly increasing use of CAD/CAM. Reorientation towards user needs, flexibility, quality, life cycle economy and ecology
2. Russia and GIS countries	3 - 5	Return to traditional building. Interest in open industrialisation in R&D. Later recovery of industrialisation assumed.
3. North Asia	3 - 6	Interest in partially open building systems has increased. Large companies are developing highly automated building concepts in Japan. Industrialisation is progressing.
4. South and Southeast Asia	1 - 6	The general trend is increasing industrialisation, but in varying ways: Through increase in manual equipment and use of small components (blocks, bricks, stones), or through prefabrication, mainly closed systems. In some countries, interest in open systems is increasing. Computer aid increasing. In several countries manual methods with equipment aid.
5. North America	4 - 6	Efficient development of advanced closed prefabricated housing units. Ecological development in focus. Advanced industrialised high rise building concepts developing further.
6. South and Central America	1 - 6	Closed local and foreign systems experiencing difficulties: more individual ways of building expected.
7. Middle East	5 - 6	Several foreign industrialised systems are continuing applications in large projects. Partial openness expected to increase.
8. Africa	1 - 4	Local applications of partially industrialised building will increase; the share of manual work will remain high. Small scale buildings predominate.

Even this rough classification shows a wide variety of stages of industrialisation in all regions. Additionally each area has its own atypical cases.

The stages of industrialisation of building in each society reflect the general phase of industrial development in the society, but also other factors such as culture and building traditions. This emerges particularly when comparing the most developed regions like Europe, North America and Japan. In Europe, where building traditions are highly respected, industrialisation is not at the level that the general knowledge and industrialisation of the society would allow. Typically larger countries seem to have a lower rate of development towards open industrialisation than do smaller ones. In fact the smaller European countries are those that lead the industrialisation process world-wide. These include Finland, Denmark, Norway and the Netherlands which all apply open industrialisation in design and production. Newly developed countries already use mass production, which means increased application of closed building concepts from foreign countries, but increasingly flexible, partly open systems are already appearing. Hopefully this will lead to the same trends as already visible in some European countries. In developing countries, both in Africa and in Asia, the industrialisation of society has not yet progressed sufficiently to allow extensive use of industrialised building. In these countries only materials production and the production of small building parts can be industrialised and used in manual site manufacture.

4.2 INDICATORS FOR ADVANCED INDUSTRIALISATION

It is interesting to try to identify common difficulties in industrialisation in different societies and working environments. The following common predictors of serious difficulties can be recognised:

1. Closed and non-flexible building systems can be applied only in large building projects, where high investments can be amortised in a single project. This is because the same building concept and design cannot be invariably applicable to different built environments and different requirements for use. This problem was reported in both the USA and from South Asian countries. This is one of the main factors to have held back industrialised building in some European countries following the period of mass production.

2. Aesthetically, functionally and technically improper design and manufacture of buildings and products.
3. Improper utilisation of the benefits of industrial production in regard to production costs, speed and quality.

In some industrialised countries the difficulties described above have been overcome. The factors leading to this kind of success can be identified as follows:

1. Post-industrialised information society.
2. Application of open, flexible systematics.
3. **Effective** application of mechanisation and automation in prefabrication.
4. **Application** of different materials and composite structures which concur with each other.
5. **Skilled** structural designers and architects who understand the peculiarities of prefabrication and industrialisation.
6. **Advanced** education of workers.
7. **Liberal** standardisation which allows new, non-traditional products.
8. **A** generally high level of industrialisation and computer application in design, manufacture and communication in the society.
9. **Tradition** of networking between companies and organisations.
10. Interest in international export of building products, engineering know-how and building projects.
11. Mass production due to a great need for buildings.

4.3 GENERAL TRENDS IN THE BUILDING MARKET AND TECHNOLOGY

A common trend in all regions is increasing use of industrially produced building products and the forward march of automation. Industrialised countries in particular are seeing a surge of industrialisation and automation in factories, in addition to which equipment and manipulators assisting manual work on site are constantly developing and traditional site machinery has increasingly built-in automation. In developing countries the main trend is increased application of manual equipment to assist manual work on site, which in these countries is still very intensive.

In most early industrialised countries building production is either diminishing or already on a low level. Thus interest in the entire system building technology is falling in countries that do not have a tradition of open system building. By comparison, in countries that already have open system building technology and infrastructure, open industrialisation is predominant and under continuous development towards increased flexibility, individuality and automation. In contrast to new building, renovation is an increasing area of production. Also in this area the manufacture of products is highly industrial, but site work is more manual and closed than in new building. Attempts to open and close industrial renovation systems and concepts have been made and some of them seem to be successful.

Another common trend in industrialised countries is the increasing impact of computers in design, manufacture, production planning, management and communication between partners of construction. The application of computer aid is very active in newly industrialised countries. In developing countries it seems to be more difficult to apply computers because of the lack of skilled people and lack of capital for investments.

A special case among industrialised countries is the former socialistic countries, where a rapid collapse of industrialised building followed in the wake of changes in the society. Prefabrication is still regarded as a symbol of socialism; the quality is considered too low and the appearance extremely monotonous. This image is expected to prevail for a long time and to be a significant barrier to effective development of industrialised building. Another factor against industrialisation is the continued focus of building and development on renovation. Once this is satisfied through radical technical and architectural changes, the recovery of industrialisation is both possible and likely.

Newly industrialised countries tend pretty much to follow the development trends of the 1960s and 1970s, which were characterised by mass production. However, the technology now applied is better developed, and more open and flexible systems are used. There is now a good opportunity for newly industrialised countries to develop their production infrastructure, organisation and management, and design and product systematics with application of open industrialisation and general industrial methods. They also have the advantage of being able to learn from the mistakes of countries which developed sooner.

In developing countries it has been recognised that building technology and housing development must follow ways which are related to local traditions as well as social, environmental and technical possibilities. Thus the development of industrialisation, although certainly expected to continue, will be slower and different from that in the developed countries.

4.4 POSSIBLE MODELS OF FUTURE OPEN INDUSTRIALISATION

Future perspectives of open industrialisation in building are promising. Several driving forces in the general evolution of societies are pulling development in an industrial direction. These include:

1. A global need for increased productivity in the building sector, in order to cut production costs in developed countries and raise production capacity in developing and newly industrialised countries. Statistics show a drastically higher growth of productivity in factories yielding building products than at site works.
2. Growing demand for guaranteed quality. This is best fulfilled in industrial and automated production both in developed and in developing countries.
3. The increasing goal towards sustainable development in building. This is best solved in industrialised countries through industrial production, which saves on energy and materials, and where wastes can be recycled better than in manual production on site. However, in the least industrialised countries an alternative solution can be a very local production in manual methods using local natural materials.
4. Loss of manual skills in industrialised countries. This continuing trend threatens the building sector unless modern methods and industrial ways of production can bring skilled workers into the building sector.
5. Reluctance of workers in industrialised countries to work in harsh and/or unsafe site conditions. This is already creating a shortage of workers at a time when unemployment is high.

Industrialised building technology is developing globally and internationally, but applications into building concepts and designs will have to be made very locally in order to fulfil the local cultural, environmental, operational and economical requirements (Sarja [19], [46], Weber [19]). Individual buildings have to be adapted to the needs of their owners and users. The industrial production can be either prefabrication in factories or mechanised and automated site production, or a combination of both.

General industrial principles and methods can be applied in building [46]. Open industrialisation can be developed as global technology, which can then be applied regionally and locally in different ways using locally and regionally produced products and materials. The general rules and models can be concretised into building concepts for defined consortia or networks of contractors and suppliers.

In order to accelerate development of international industrialised building technology, research should be directed at further systematisation of performance concepts and of modularised system rules in product systems, organisation systems and information systems. The systematics should be presented as model designs, alternative organisational models and applied product data models. It is important to identify and analyse productivity factors and use the results to develop methods for improving productivity.

For implementation of the results into practice, companies have to develop long term strategic development projects over 5 to 10 year periods. These should include overall business strategy, product and method strategy and information system strategy.

The strategies should be followed by development of business networks and consortia and definition of the building concepts for them. These building concepts can be described and tested through model designs and experimental building projects. Development can be divided into three phases:

1. Implementation with current products using new organisational procedures, computer applications and design solutions.
2. Implementation by means of partial product and method development and organisational changes.
3. Implementation exploiting the entire potential for the development.

In all subareas, an important task is the transfer of generic principles and models from other fields of technologies into building, and participation in applied technical research in the core areas of industrial technologies. Core areas include e.g. the STEP systematics of product modelling and several ISO standardisation works regarding technical specification systematics.

CIB is providing information on open industrialisation from several Working Commissions, whereas W24 "Open Industrialisation in Building" focuses on the defined core areas and on organising meetings and symposia for general discussion. The U.S. Civil Engineering Research Foundation has activated a research agenda "Engineering and Construction for Sustainable Development", which includes the prospectus "Exploring the International Use of System Modularity for Construction Facilities" [12],[13]. Researchers and companies are invited to actively participate in these activities.

5

Further reading

1. Adhikari, A. P., A Proposal for Appropriate Standards for Low Income Housing for Kathmandu, Nepal: Guidelines and Samples. International Journal for Housing Science and its Applications. Vol.18 No.3 1994. Pp. 167 - 202.
2. Adler, P. CIB-W24: Open industrialization in building , State of the Art in Sweden. 1996. 9 p.+ app. 6 p.
3. Adler, P. Open Industrialization in Building, Case Studies. In Report of CIB W24 meeting, November 6-7, 1995. The University of Reading.
4. Adler, P., Sarja, A. Questionnaire form for the state of the art - analysis of industrialised building. Sweden, apartment buildings. Open industrialization in building. 1996. 3 p.
5. Ajibola, K., Olubodun, O.F. Appraisal of low cost housing design - case study of a federal government housing typology. International Journal for Housing Science and its Applications Vol. 14. No 2. 1990. Pp. 91 - 105.
6. Akarakiri, J.B. An analysis of pre-fabricated trussed rafters in Nigeria. International Journal for Housing Science and its Applications. Vol. 16. No 4. 1992. Pp. 253 - 263.
7. Al-Khaiat, H. Impact of the use of precast building systems in Kuwait housing projects. International Journal for Housing Science and its Applications. Vol. 15. No 2. 1991. Pp. 133 - 149.
8. Al-Khaiat, H., Qaddumi, N. Technical views on the use of prefabricated building systems in Kuwait housing projects. International Journal for Housing Science and its Applications. Vol. 13. No 3. 1989. Pp. 243 - 250.
9. Bertelsen, N.H. The Innovation of the Building Process in Denmark. Building research institut, Denmark. 3 p.
10. Blachère, G. A reasoned introduction to the work of standardization for the development of the use of components and for the "open system". Library Translation 1785. Building Research Establishment. Department of the Environment.

Building Research Establishment, Depart of the Environment. September 1973. 37 p.

11. Brown, G.Z. Industrialization of Single Family Housing in the United States. Energy Studies in Buildings Laboratory Center for Housing Innovation. University of Oregon. 1996. 8 p.

12. CERF Civil Engineering Research Foundation. Report 96-5016.E Creating the 21st Century through Innovation. Engineering and Construction for Sustainable Development. Executive Report. Washington, D.C. 1996. 57 p.

13. CERF Civil Engineering Research Foundation. Report 96-5016.T, 1996. Construction Industry Research Prospectuses for the 21st Century. Technical Report. Engineering and Construction for Sustainable Development. Washington, D.C. 1996. 130 p.

14. Cuperus, Y. Ype in China. Reisaantekeningen van een bezoek aan de Volksrepubliek China van 16 tot oktober 1994. Teknische Universiteit Delft OBOM Onderzoeksgroep. 35 p.

15. De Troyer, F. Open Industrialization in building, definition of objectives, concepts and the role of dimensional co-ordination. CIB W24 Meeting Helsinki, 22 May 1995.

16. De Troyer, F. Principles of modular co-ordination in building. Draft for revision by IMG/CIB W24. Katholieke Universiteit Leuven Departement ASRO Kasteel van Arenberg B-3001 Heverlee, Belgium 30 June 1991.

17. Enkovaara, E., Salmi, M. & Sarja, A. Ratas project: computer aided design for construction. Helsinki 1988, Building Book Ltd. 57 p. + app.

18. Erdely, A., Zaicescu, D.C., Ionescu, A. Construction system with space frame units (S.F.U) International Journal for Housing Science and its Applications. Vol. 14. No 2. 1990. pp. 83 - 90.

19. European Communities Commission. Stuttgart February 21-23, 1990. Open industrialization a solution for building modernization. Conference proceedings.
   Davidson, C.H. Open industrialization: Technical and organizational prerequisites. Pp. 4-17 -4-26.
   Hannus, M., Computer aided design of component based building. Pp. 6-1 - 6-10
   Huete, R. Proposal to improve the contribution of architecture schools to open industrialization. 4-53 - 4-62.
   Lugez, J. A means of industrialization: Construction by components pp. 4-1 - 4-16.

Moulet, J. Construction, dimensional and computerized coordination, production of structural components industrial elements, consequences on production plants. Pp. 2-1 - 2-9.

Rhoul, B., Salem, R. Juxtaposition of GCinq approach with the reality in field. Pp. 6-83 - 6-112.

Sarja, A., A new generation of open building based on hierarchical modulation. Pp. 6-21 - 6-30.

Weber, J., The advantage of open industrialisation and dimensional coordination for developing countries and for the opening of regional market. P.p. 7-11 - 7-22.

20. Ezz Al Din, M.A. Costs of overdesign of public housing projects in Kuwait. International Journal for Housing Science and its Applications. Vol. 14. No 3. 1990. Pp. 181 - 196.

21. Feodorow, E. Housing policy and particularities in solving the housing problem in the USSR. International Journal for Housing Science and its Applications. Vol. 13. No 1. 1989. Pp. 13 - 21.

22. Friedrich, R., Beckmann, Utz. Industrialized Precast Construction Methods for the Construction of Housing and Public Buildings in the GDR. Betonwerk + Fertigteil-Technik 6/1990. Pp. 45 - 49.

23. Fukao, S., Matsumura, S. Status of Open Industrialisation in building Country: Japan. The report for CIB W24. Tokio 1996. 16 p.

24. Gibert, M. Industrialization of urban housing construction in India: A survey of suitable shapes. International Journal for Housing Science and its Applications. Vol. 16. No 3. 1992. Pp. 179 - 187.

25. Hetherington, W.T. The Canadian Response to Open systems for Building Automation. ASHRAE Journal November 1994. Pp. 40 - 44.

26. Hurez, M., Moulet, J. G5 - Software to Develop Open Industrialization of the Building Industry. Betonwerk + Fertigteil-Technik. 3/1994. Pp. 85 - 88.

27. Industrialised building systems in Asia. Low-Cost Building Materials Technologies and Construction Systems Monograph Series Number 3 June 1988. Regional Network in Asia for Low-Cost building Materials Technologies and Construction Systems (DP/RAS/82/012). A project Funded by the United Nations Development Programme (UNDP) and Executed by the United Nations Industrial Development Organization (UNIDO). Manila 1988. 236 p.

Aureus-Eugenio, V. Philippines, Adaptation of industrialized building systems in the implementation of low-cost housing programs in the Philippines. Pp.167 - 204.

Chaimungkalanon, S. Thailand, Semi-industrialized building system for housing in Thailand. Pp. 221 - 236.

Commandante, A.S. Philippines, Industrialized housing systems. Pp. 155 - 165.

Ediriweera, S.R. Sri Lanka, Semi-industrialized construction technologies. Pp. 205 - 220.

Faruq, M. Afghanistan, State-of-the-art and problems related to adaption of industrialized building systems in the democratic republic of Afghanistan. Pp. 1 - 30.

Goupan, L., Yan, Z. China, Development of industrialized housing systems in China, its status and prospect. Pp. 67 - 75.

Maher, A., Hasan, S.T. Pakistan, Industrialized systems for housing. Pp. 143 - 153.

Partadinata, D.S. Indonesia, The state-of-the-art and problems related to the adaptation of industrialized building system in Indonesia. Pp. 77 - 132.

Roy, R.K. Bangladesh, State-of-the-art of building materials and construction technology in Bangladesh. Pp. 31 - 66.

Swee, H.P. Malaysia, A review of industrialized building systems. Pp. 133 - 141.

28. Innocent, A. Mitigating Nigeria's housing shortage with brick-based housing. International Journal for Housing Science and its Applications. Vol. 14. No 3. 1990. Pp. 219 - 231.

29. Kendall, S. Developments Toward Open Building in Japan. Open House international v. 21 no 3 1996. P 41.

30. Koncz, T. New Technology Spurs Market for Large Panel Precast Concrete Buildings. PCI Journal Jan-Feb 1995. Pp. 30 - 42.

31. Korea National Housing Corporation, Housing Research Institute. Principles and Recommendation for the Modular co-ordination building in Korea. 1996.

32. Lahdenperä, P. Open Industrialization in Building- Framework for the comprehensive development. The CIB working commission W24 'Open Industrialization in Building', May 22-23. Espoo 1995. 8 p.

33. Lahdenperä, P. Reorganizing the building process, The holistic approach. VTT Publications 258. Technical Research Centre of Finland. Espoo 1995. 210 p.+ app. 7 p.

34. Larsson, N. Green Development Corporations. A proposed framework for an economically attractive and environmentally sustainable form of urban development. Canmet, Canada. October 10.1995. Unpublished paper. 13 p.

35. Leppänen, P. & Sarja A. Preliminary models for the computer aided building design and construction management process. Espoo 1987, Technical Research Centre of Finland. Research Notes 606. 47 p.

36. Lyall, S. CLASP. Architects'journal 7/1996, pp. i - vii.

37. Misra, S.K., Sarja, A. Questionnaire form for the state of the art - analysis of industrialised building. India, apartment buildings. CIB W24: Open industrialization in building. 1996. 6 p.

38. Mohan, S., Johnston, D.W., Shoemaker, W.L. Residential Construction: Research Agenda for the 1990's and beyond. International Journal for Housing Science and its Applications. 15.3.1991. Pp. 211 - 228.

39. Open Industrialisation in building . Report of CIB W24 meeting. The University of Reading, department of Construction Management & Engineering. November 6-7, 1995.

40. Prieto-Portar, L.A. An advance in 3-dimensional modular housing. International Journal for Housing Science and its Applications. Vol. 18. No 4. 1994. Pp. 203 - 215.

41. Prieto-Portar, L.A. "Funding 525,000 units of public housing in Argentina". International Journal for Housing Science and its Applications. Vol 12. No 1, 1988. Pp. 1 - 16.

42. Principles and Recommendation for the Modular co-ordination building in Korea. Korea National Housing Corporation, Housing Research Institute. 1996.

43. Ramamurthy, K.N. Partial industrialized system for mass housing. International Journal for Housing Science and its Applications. Vol. 12. No 1. 1988. Pp. 17 - 21.

44. Reddy, P.R., Lefebvre, B. Rural housing and perception of inhabitants - case study of an indian village. International Journal for Housing Science and its Applications. Vol. 17. No 1. 1993. Pp. 49 - 55.

45. Russell, B. Building systems, industrialization, and architecture. Norwich, England 1981. About 700 p.

46. Sarja, A. & Hannus, M. Modular systematics for the industrialized building.VTT publications 238. Technical research centre of Finland, Espoo 1995. 216 p.

47. Sarja, A. Towards the advanced industrialized construction technique of the future. Betonwerk + Fertigteil-Technik 4/1987. Pp. 236 - 239.
48. Sawada, S. "Design and Build" Projects by Japanese General Contractors: SMART System project and Urban Development Project. Bauforum Berlin 1996, September 19-21, Former Berlin Staatsratsgebäude Session: Integration in the building process- A path into future, September 19, afternoon. 5 p.
49. Sawada, S. Open building Development in Japan. 6 p.
50. Viscomi, B.V., Michalerya, W.D., Lu, L-W. Automated construction in the ATLSS integrated building systems. Automation in Construction 3(1994). Pp. 35 - 43.
51. Warzawski, A. Status of an open industrialization in Israel.1996. 3 p.
52. Warzawski, A. Industrialization and Robotics in Building. A Managerial Approach. National Building Research Institute Technion-Israel Institute of Technology. 1990 New York. 466 p.
53. Zletni, B.R. The development of housing programs in Libya. Housing Science Vol. 17. No 1. 1993. Pp. 19 - 30.
54. Michael, Duncan, Concepts of composite construction - Mutatis Mutandis. In: Composite Construction - Conventional and Innovative. Conference report, International Conference, Innsbruck, Austria, September 16-18, IABSE. Pp. 19 - 28.
55. Vukovic`, Svetlana, Industrialized building in Yugoslavia. CIB W 24 International Seminar on Industrialization in Building: Present State and Future Trends. Haifa, Israel, November 4-7, 1997. Pre-Seminar Proceedings. TECHNION, national Building Research Institute. Pp. A4 - A11.
56. Dahan, David & Warzawski, Abraham, Industrialized building in Israel. CIB W 24 International Seminar on Industrialization in Building: Present State and Future Trends. Haifa, Israel, November 4-7, 1997. Pre-Seminar Proceedings. TECHNION, national Building Research Institute. Pp. A23 - A32.
57. Ting_Ya Hsieh, Industrialised building in Taiwan. CIB W 24 International Seminar on Industrialization in Building: Present State and Future Trends. Haifa, Israel, November 4-7, 1997. Pre-Seminar Proceedings. TECHNION, national Building Research Institute. Pp. A12 - A22.

Part 2

Selected personal visions

Selected personal visions

1

In search of a better way of building

Peter Adler

D.Tech. (Sweden)

1.1 A RETROSPECTIVE INTRODUCTION

Experience with the utilisation of the first generation of industrialised building systems for mass-production building and the difficulties which followed the building industries' changeover to small-scale production, contributed to some scepticism towards industrialised building methods. Mass-production building, its forms of procurement. its building techniques and manufacture were not developed to cater for small scale production.

In most European countries, when the objectives of governmental housing policies were within reach, the interest for technical development decreased. Technical advances which characterised mass construction in housing after World War II have not been followed up. Research and development in building industry has been quite strictly directed towards internal technical and management questions and less attention has been paid to the interaction of social prerequisites and the technical solutions that have been developed.

Serious social and structural consequences occurred in the housing sector as well as problems in other sectors. An example is the extensive publicity tracing the "sick-building-syndrome", a phenomenon which arose in most countries which were extensively utilising the first generation of building systems. Obviously a new approach was needed for tackling the questions of industrialisation in building. During a period of transition from "closed" to "open" building systems in the nineteen seventies and eighties, a systematically developed and more open approach has been tried in Denmark and Finland. This approach contributed to the development of an advanced component-building technology, more known under the name of 'catalogue building' - which became one of the main characteristics of the second, the open, generation of building systems.

Rational, industrialised building with prefabricated components presupposed co-ordination of sizes, performances and joint characteristics. Standardised rules for modular co-ordination, performance analysis and jointing of components, proved to be vital instruments in the development of the component technology. The increased range of components created, in its turn, a demand for simple and easily understood technical literature and planning guides. Experience proved that actual, systematised and open product information were crucial for the implementation of prefabricated building components and building parts. This kind of information enabled the performances of the building products to be assessed at the outset of the building process. Systematic surveys of available components and building parts on the market facilitated their comparison and interchangeability during procurement procedures and granted variation in design, thanks to an increased freedom of choice regarding materials, performance and technical solutions.

Although standardisation and open information became doubtlessly two necessary preconditions for the coming into existence of any national or regional component market, these preconditions were however far from being sufficient to guarantee the functioning of an open component market as well. Experience showed that other preconditions, tied to the general acceptance of the rules of co-ordination and of the open approach for development in building, have to be spread widely and practiced appropriately of all the parties in the building process, if the advantages of prefabrication should be utilised.

What the open system approach clearly demonstrated was a substantially higher degree of adaptability to the variety of local conditions for building than that which previous, closed approaches could exhibit. However, the gradual realisation of governmental housing programmes later on, with successively reduced subsidies and other restrictive measures as well as the demands for more attention to the environment and energy conservation - created a totally new demand-picture for housing provision and for the development of building techniques.

1.2 CHANGED PRECONDITIONS AND NEW CHALLENGES

Some fifty years after the end of World War II, the quantitative side of the provision of housing can be considered as substantially solved in almost all European countries. With the age of major government investment programmes left behind, the building industry has lost its traditional role of needing large labour forces in most national employment markets. The advantages of industrialised building methods must be utilised in quite different contexts.

One of the "new" tasks of a building industry is certainly to safeguard the quality of the existing housing stock. Renovation, rebuilding and improvement of the existing stock can hardly be carried through with traditionally industrialised building methods (closed systems), but rather with industrially manufactured building parts and components. Another, though not so new, task of the building industry is to increase productivity and reduce building costs. This is, certainly, possible with a reasonable utilisation of industrialised methods of production, but, built volume per hour, the traditional yardstick of productivity is insufficient to satisfy the new demands. Other and more complex forms of rationality, comprising inter-alia good function, adaptability of building methods and compatibility of building components are also required and have to be taken into account when considering the productivity of industrialised building methods.

Today, there is a completely changed scene in the home market of almost all "national" building industries in Europe. The building industries have to compete with other industries for and with qualified personnel, with the quality of the built environment and with a continuous product development in accord with the demands of the public. The environmental performance of all industries - including the building industries - grows more and more important as a competitive factor and in the evaluation of the adaptability of technical solutions to local and regional conditions.

In the wake of repeated raw material crises and environmental disasters, the connection is seen between the standard of living and the limits of natural resources. This realisation lays the foundations of a new and complex pattern of requirement for design and building by providing new criteria for the evaluation of performance.

With the development of ecology~based knowledge and technology within planning and construction. building has to face new challenges. The choice of starting point of this new knowledge is: the human being, the local situation, resource management and an ecocycle approach - aimed at preserving and developing the diversity of human habitat.

Resource management and adoption of an ecocycle approach generates entirely new conditions for planning, building and technical development, which is traditionally based on the view of unlimited and relatively cheep access to energy and materials. Current discussions on the matter indicate, however, quite a new direction for theoretical development, claiming an account of long~term effects and sustainability of the design as well as of industrial manufacture and of building. Moreover and predominantly: there is the enormous housing shortage in the Third World which, in my opinion, should be considered as the virtual challenge for any kind of industrialisation in building.

1.3 INDUSTRIALISED BUILDING - FOR WHOM, BY WHOM AND HOW?

In both industrialised and developing countries there are shortcomings in the connection between building technology and housing function. The opportunities provided by building technology are not utilised to satisfy the needs of everyday life. For instance, the provision of housing has dramatically deteriorated in some developing countries and the absence of appropriate technology contributes to the problems. Other examples are industrialised countries failing to develop an appropriate technology necessary for a provision of housing without unreasonable expenditure of resources.

Most experience of industrialised building has so far been gained in countries with well developed economies and an established industrial sector. Where industrialised building has been introduced at all in Third World countries, the prototype has been taken from industrialised countries, often with the aim of leaping over stages in development in order to more quickly reach production capacity and quality perceived as being unattainable with traditional local methods. Commonplace examples from developing countries are simple but heavy panel systems which do not work because the road system in incapable of taking the loads involved, or of systems demanding a high level of accuracy which cannot be made by unskilled labour in co-operative organisations. These sort of failures which have been noted have hampered a constructive analysis of cause and effect and thus prevented a systematic adaptation of technique and forms of production to the locally prevailing conditions.

The result has often been a return to the old methods and an undiminished scepticism toward change.

This does not necessarily mean that the industrialised countries' building technology is incompatible with conditions in the Third World. But, instead of transferring technicians, industrialised countries should be developing special technical solutions for, or assisting the development in, the countries where the need is greatest. The absence of adequate building technology in developing countries should reasonably be seen also against the background of the confusion which exists regarding the difference between aid and export. The long-term aim of aid - in the context of the developing country - to contribute to the establishment of autonomously functioning units of production, administration and other organised work, does not always tally with the (admittedly unspoken) aims of export, and also its constant sequel: to open new markets, based on long-term dependence.

However depressing these experiences can be I find it equally encouraging to remember that traditional building which is often praised for its wealth of environmental qualities and differentiation, is often built up of a very small number of formal components within very tight economic and technical limits. From this, it is possible to draw the following conclusions:

- building techniques comprising a very limited assortment of components, need not necessarily lead to monotony and aesthetic poverty,
- sophisticated building techniques can become public property if the technology is attuned to prevailing conditions,
- an analytical instrument is needed that embraces a wide range of conditions for industrialised building, with the purpose to analyse and evaluate the appropriateness of technical solutions.

The aim is to make it possible to overcome the scepticism and also, perhaps more importantly, to make possible changes in technical development in a direction that would turn industrialisation into an effective means of tackling the housing shortage in the Third World.

Excursus   We will, of course, always remember the period of rebuilding Europe after World War II as the golden age of building systems - and of system builders. This was, however, a half a century ago, when Europe and the World had to be armoured for other tasks. In present

103

times, though, after the "post-wall euphoria", we had to bear witness to the unprecedented destructive power of "advanced technology" destroying people and human habitats in Bosnia, in Chechnya and elsewhere in Africa and Asia. I just wonder if we are capable of supplying those who survived, as well, with an appropriate technology for the rebuilding of their communities?

1.4 CIB W24: OPEN INDUSTRIALISATION IN BUILDING

The preceding presentation gives a good reason to call into question whether the development of technology can bridge the problems created by changes in the economic and political conditions for housing construction.

Outside the building sector new technologies are gaining an ever stronger foothold. Information technology, automatization and robotization in the manufacturing industry stand out today as powerful challenges and tools, offering opportunities which could also be put to good use in building. A new requirement pattern emphasises, with increasing clarity, the requirements of; the users, of building maintenance and of ecology, side by side with the technical, economic and production possibilities and constraints. Research and development activities carried out in, for instance, Denmark and Finland, have established some basic principles for the development of the next generation of technical solutions in building. The contours of open industrialisation in building are beginning to take shape. For me, today, open industrialisation in building means: integrated and systematic exchange and implementation of products, services and information between the parties of the building process nationally; and internationally between countries and regions which have major differences in the structure of their building industry.

This is a definitively new way of looking at building. It is a new approach, with the aim to set it free from the restrictions of short-sighted thinking in closed systems and open it towards the environment in which it exists and towards its innate creative potentials with the integration of industrialised manufacture, rationalised craftsman production and automated handling of information.

As current national developments are mostly carried out with somewhat fluctuating preconditions, the questions touching their co-ordination and any possible collaboration between them, have a steadily increasing importance. These questions are dealt with, amongst others, within CIB:s working commission W24: Open industrialisation in building.

1.5 ASSESSMENT OF TECHNICAL SOLUTIONS IN BUILDING

Today, when the internationalisation of building techniques and building products is becoming increasingly tangible, it is important that the local and regional characteristics in the physical environment are not brushed aside by superficial international design fashions. For this reason, a systematic description of situational preconditions and restrictions can be an important instrument in the adaptation of building processes to the physical and societal environments in which buildings have to be erected. In the following, my point of departure is that the environmental performance of building techniques should be considered as proportional to the techniques' adaptability to the requirements specified on every particular site and to their adaptability to changes of requirements over time. This seems to be a plausible reason why the systematic introduction of "non-technical" requirements in technical research and development and in routine housing construction should be considered as effective means for a real modernisation of the building industry.

The experience of the Nordic countries, and of others, shows that one necessary condition for the development of, in the widest sense, a good building technique is to systematically evaluate the existing technique, taking as a starting point the requirements which it should fulfil.

Industrialised building techniques, like any other techniques guided by internal objectives are certainly capable of repeating achievements from one situation to another, but without an inherent adaptability to the varieties and to the changes in the surrounding environment these achievements will in course of time reveal an increasing gap between the actual requirements stated and the factual performances achieved. Such techniques will very soon fall into disuse and have to be replaced by contemporary and more adaptable solutions. Thus, the capability of elaborating and assimilating various and changing exterior requirements from its environment, remains the fundamental precondition for an acceptable implementation of any technical system. It is also necessary for the satisfactory performance of a building system. The question is, as I see it, not if but how social, ecological, political and other changing criteria can be added to a requirement pattern hitherto dominated by narrow technical and economic criteria.

A possible starting point for the elaboration of an analytical instrument for description and assessment of technical solutions in building is to consider building as a sub-system in the wider context of society's structure and needs. This structure embraces not only the technical conditions of building, but also the social, cultural, economic and other segments of the society, of which building is both a prerequisite and a result. It is by analysis of this complex structure that the opportunities and the constraints for technical development can be described and thus provide a basis for development. It is essential that the analytical instrument comprises a matrix that describes both the building techniques and the situational factors generating specific demands.

The description of building techniques covers supporting structure and equipment systems at a number of levels required, for instance, at the levels of: neighbourhood, building type and dwelling plan. In this case the principal level of the description is the building level, at which the requirements for various types of (residential) buildings are dealt with. The main characteristics of the technical solutions are, thus, arranged along the vertical axis of the matrix.

Along the horizontal axis will then be arranged the performances of the technical solutions, related to internal and external factors of importance in any particular site. Here the meaning of internal and external rationality has to be introduced and explained:

- as internal rationality of technical solutions, I consider the correspondence of their performances with criteria regarding, for instance, their adaptability, compatibility, exchangeability and others.
- as external rationality of technical solutions, I consider the correspondence of their performances with criteria which are found in complex background factors, such as: cultural and social systems, infrastructure, the level of technical development and so on.

However, any arrangement of a set of criteria for assessment of technical solutions has to take into account certain overlying conditions, as for instance:

- The demands for increased adaptability of production methods to situational preconditions. These demands point at the external rationality factors as being crucial for any adequate assessment of the environmental performance of building techniques to be applied.
- Assessment procedures which primarily take note of conventional technical aspects, will probably work in countries with a sophisticated

industrial structure which has an infrastructure and building industry in accord with the general level of development. But what is needed is an instrument for assessment of technical solutions for housing construction which is also applicable in developing countries.

- The increased impact of "the ecological imperative" on everyday life, underlines the adoption of resource management and an ecocycle approach as the necessary starting points of theoretical and practical technical development in industrialised building. This means that the development of building products and planning methods have a vital position for the elaboration and introduction of the principles for environmental sustainability as well as for the methods of analysing life-cycle and recycling procedures for buildings, components and materials.

- The present stage of development in industrialised building, which can be characterised as a transitional period between the second and the third generation of systems development. One of the distinguishing features of the period is the transformation from more or less mechanised manufacture to a partially or entirely robotized production of building parts and components.

- The trend to "globalisation" in the building industry which has occurred in the wake of an intensified internationalisation after the end of the cold war. The obligations of the building industry as well as of those participating in, or contributing to, its development must not diminish - or what worse disappear - within the "globalizing" trend. On the contrary. Chasing and penetrating new markets means a preparedness to meet increased demands for variety in expression and for adaptability to ever changing social, cultural, ecological and other non-technical preconditions for construction.

- In every society there exists a certain dynamism which if properly channelled, contains a range of opportunities for improvement of the conditions of life. Building can provide the driving force in this kind of positive development, amongst other things, by supporting constructive social structures and by bringing forth new ideas in social and cultural fields. This, in my eyes desirable, development presupposes an ability to take note and interpret latent opportunities for change. To satisfy these extremely complex demands will be the real challenge of a truly industrialised building industry.

These are, roughly taken, the points of departure for a coming project covering a number of case studies of hitherto developed open building methods and techniques. The aim is to identify, describe and exemplify the openness of the accompanying technical solutions. For this purpose a comprehensive and open ended system for description and assessment has to be developed - a system which can be added to and varied, to meet defined aims regarding areas of analysis. The project is on the way to being established at the Royal Institute of Technology, Stockholm, if and when proper financial support and other fringe conditions can be arranged.

1.6 IN SEARCH OF NEW FORMS OF PROCUREMENT

The technical development of the mass-housing period was based on competing closed systems. Instead of increasing the choice between competing solutions, clients and designers became dependent on proprietary products. This development was hardly beneficial even during the building of the giant schemes of mass-housing. Nor did it provide favourable conditions for renovation, rebuilding and improvement of the existing housing stock later on. One of the lessons we in the Nordic countries have learnt from the first period of industrialisation in building is that industrialisation must not be carried through entirely with closed systems, just nominally competing among themselves. Danish, Finnish and Swedish experiences can clearly demonstrate this.

The forms of procurement developed and used in mass-housing construction were fairly convenient due to the highly specialised roles of the participants. Some forms of procurement were very effective for large-scale schemes on green field sites, inspiring extensive and continuous research and development exercises, as long as the state guaranteed the necessary financial, organisational and administrative means for the production. Seen from another standpoint, one may say that in mass-housing, technical development has been promoted by every form of procurement.

Although the building systems competed with isolated products, the essential item for the systems' environmental performance - the adaptation of the process of construction itself - was, and seemingly still is, untouchable. What I have in mind here is that the construction process itself (that is; the preparation, the manufacturing, the assembly of components, the organisation of the building site, and so on), should be the product offered to the client, rather than the multiplicity of more or less controllable and all too often incompatible products on the market.

A comprehensive approach which, has some possibilities of reaching the objective - an increased adaptability of industrialised construction methods - appears to be the opening-up of the building process towards its innate creative potential and towards the environment in which it exists.

Opening-up the process "inwards" means taking care and making sense of the ability and creativity of the people involved in manufacturing, construction and in other phases of building. The opening-up "outwards" mean to increasing the receptivity and openness of industrialised methods towards the variety and changes of requirements in the surrounding world. The opening-up of the building process is presumably realisable through the gradual introduction of some new forms of procurement with aims as:

- to assure the influence and sustain the interests of the users in the successive phases of the building process,
- to enable the increase in skill and competence of the labour involved and to make sense of people's creativity and inventiveness,
- to strengthen the relations between the client, his consultants and the users - as a united counterpart to the contractor,
- to support, within reasonable limits, the adaptability of the building process to local conditions,
- to obtain the independence of material producers and suppliers from the influence of general contractors.
- to promote open competition between the contractors in the market,
- to stimulate technical development, tried and tested within small-scale housing projects,
- to make possible continuous technical development exercises and feed-back studies within the limits of discontinuous production.

1.7 THE ROLE OF THE ARCHITECT

Experience gained with industrialised building underlines that rational design with prefabricated components presupposes firstly, that the manufacturer of components has published exhaustive and systematic technical information containing, inter-alia, satisfactory solutions for the jointing of the components in various situations and secondly, that the designer has developed a special technique which makes use of the possible gains in effectiveness.

Here, the role of the skilled architect cannot be overestimated, as the appropriate use of the technical solutions in building depends in many respects on the architects knowledge and role in the building process.

Technical regulations, available manuals and actual product information are useful sources, but the architect's understanding and hard-earned practical knowledge of functional and technical requirements and of the conditions for building production are, on the end of the day, vital for the desired outcome.

The role of the architect in the process of building is twofold. On one hand, the role is articulated in planning and design, when interpreting the users needs and formalising them in drawings, specifications and other documents. On the other hand, the architect's role is articulated when analysing the consequences of the chosen technical solutions, translating their performances into terms of human behaviour and experience and into relations between quality and costs of the chosen solutions.

Today, when building techniques are in all essential respects still affected by the methods of large-scale building, the development of a socially desirable and a technically feasible housing construction technique should not be treated as an isolated question for technicians. Socially and functionally oriented architects, sociologists, planners and all those whose work touches on housing should take part in the shaping of an appropriate building technique which will cope with the new requirements.

My view is that the successful development and positive change in construction techniques for housing have to be carried out in a permanent interaction with the demands of the users. We need to recognise that the development and the implementation of technical solutions - especially of those in the building trade - are in an ever increasing degree a part of societal processes with consequences far beyond the sphere of technique and of its practice.

If houses, buildings, the built environment and human settlements of the future, built with any new, or old, building techniques, are to be considered as something more than a mere reflection of the applied techniques and of its practice, it would be beneficial for all parties of the building process if skilled and socially engaged architects would notify their intention to take part in the development of building techniques.

The new way of looking at building which is gradually emerging, aims ultimately at the liberation of the field of action from the rigidity and the curtailments of a doctrinaire and misunderstood systems thinking to open the field for a truly user-oriented and environmentally friendly construction technique for housing. The socially oriented and skilled architects endeavour in this project is indispensable. The question is not why, but rather how to participate.

In industrialised countries and in developing countries as well, the education of architects (and of other traditionally schooled specialists engaged in building) must in some essential respects be reoriented to consider the demands and prerequisites "in the rest of the World". If there shall be any use of architects in the future, their education has to be completed with a range of other subjects equally important in the First as well as in the Second and the Third World.

1.8 RESTRUCTURING IS NOT ENOUGH

The building industry in most industrialised countries has in the past been a domestic industry, more or less shielded from forces of international competition and internationalisation that in most other industrial sectors has already taken place. There is, however, an ongoing restructuring within the building and the building materials industry, for which a shrinking home market appears to be the main "driving force". When it comes about, in approximately twenty years or so, this trend will perhaps end with a very few but very large building and building materials producing companies dominating a number of regional markets in Europe.

Contrary to this tendency, within limited home markets, small contractors will probably emerge, to complete the local housing projects and reconstruction works. For these small firms as well as for the "globalised giants" their adaptability to the local markets' specific conditions and arrangements is - and appears to remain - the crucial factor for their survival. A modernisation of the building industry will, in all probability, take place in the not too distant future. It seems reasonable to suppose that it will be carried through by highly skilled individuals, professional firms and organisations in international collaboration, irrespectively of their size and with a considerable freedom of action within given limits. Advanced computer-aids, systematised quality-assurance and organisational renewal, appears to be a reasonable means for examination, in order to find acceptable alternatives to the present situation.

Nevertheless, one of the decisive factors for the breakthrough of the idea of open industrialisation in building is, undoubtedly, the existing structure of the 'national' building industry. By this I understand the constellations of, and the relations between large, medium-size and small material producing, consulting and constructing firms, constituting the 'national' building market. With good reason one may wonder what possibilities these "local structures" have, or do not have, for the adoption of the idea of open industrialisation in building.

From my own experience, and I believe from that of others' too, I may conclude with the apparent statement: steering the technical development in building by giant enterprises, or by the state alone, neither guarantees anything nor does it tally so well with the idea of a free market. So, I think, in order to really develop and modernise building techniques, we have to strengthen the new road, the one of open industrialisation, which hopefully will bring us to Rome.

Essential here will be to imagine the new road as one which will be open for everybody - as open industrialisation in building has to take place in regions and countries with rather different structures of the building trade. This might be a rather good reason why we have to consider open industrialisation in building as an open-ended process, adaptable to the variety of ever changing local and regional conditions for development. This capability for adaptation will be the foremost and the true warrant for the development of open industrialisation itself.

REFERENCES

1. Adler. P, 1985, (1990 in English), Two chapters: New-build and Develop appropriate technology, in Housing research and design in Sweden (ed S, Thiberg), Swedish council for building research, publ no: D13:1990, Stockholm.
2. Adler, P and Thiberg. S, 1986, Expanded method of evaluation for industrialised building (paper submitted to CIB W19:s session in Budapest).
3. Sarja, A, 1989, Principles and solutions of the new system building technology (TAT), VTT research reports 662, Espoo.
4. Malbert, B, 1992, (1994 in English), Ecology-based planning and construction in Sweden, Swedish council for building research, D3:1994, Stockholm.
5. Sigurdson, J, 1994, Internationalisation: a new basis for quality in building research, Swedish council for building research, BVN 1994:1, Stockholm.
6. Henricson, E. and Jacobsson, S. 1994, Strategisk studie av den europeiska byggsektorn (Analysis and conclusions from Swedish standpoints of the "Atkins-report", in Swedish), Swedish council for building research, report no R31:1994, Stockholm.

7.   Adler, P, 1995, Bostadsbyggande - på väg mot öppen industrialisering, (Housing construction - towards open industrialisation, in Swedish), AB Svensk Byggtjänst Förlag, Stockholm.

8.   Adler, P, 1995, Open industrialisation in building - case studies, (a project proposal submitted to CIB W24 session in Reading).

9.   Adler, P, 1996, Open industrialisation in building - a state of the art in Sweden (report submitted to CIB W24).

10. Adler, P, 1997, Some aspects of industrialised building in Nordic countries - Swedish, Danish and Finnish experiences (paper submitted to the CIB W24 session in Haifa).

2

Measuring productivity and quality on the building site in Denmark

Niels Haldor Bertelsen
(Denmark)

The people of the Danish building construction sector are well-educated with a high standard resulting in well-planned houses and apartments of good design and high quality. The sector focuses primarily on the home market, but in recent years more and more firms look towards markets further away.

The activities of the sector have undergone great changes, an ever greater part being spent on urban renewal and renovation so that new building and renovation are of equal size (fig. 2.1.).

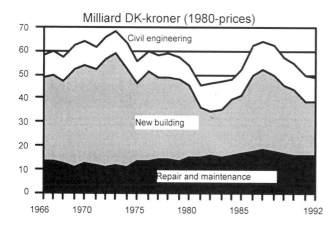

Fig. 2.1. Accumulated production value of the Danish building and construction sector.

As in other countries the sector is heavily reliant on and sensitive to economic trends. It is therefore of great importance to the national economy. In Denmark families spend 20% of their income on the home and the total building and construction sector including building products constitutes almost an equal share of the national economy.

Although the sector is an important national economic factor, it is characterised by low industrialisation and a constant or decreasing productivity. A well-known problem in the industrialised countries, and a problem which Denmark has tried to remedy for a long time.

In order to solve this problem, public efforts in Denmark during the last 25 years have been concentrated on the following areas concerning development and legal measures:

- Energy-saving in connection with room heating and energy
- Quality assurance in the execution of building
- Information technology and EDT applied within the sector
- Indoor climate, environmental control and ecology within the entire sector
- Urban renewal and renovation of buildings in urban areas (Project Renovat.)
- Development of subsidized housing and new forms of cooperation within the sector (Process and Product Development within building)
- Architecture and aesthetics of Danish house building
- Development of new types of buildings with different profiles of properties

The whole building and construction sector has taken an active part in these initiatives. The field of industry and materials production has seen an increase in productivity and a strong refinement of the products. Today the products cover more functions and require less finishing on the building site, but there is a great variation within the various production areas.

In the planning and building field many attempts have been made at development but with little impact on the industrialisation and productivity of building parts or on the sector as a whole. In this part of the sector much weight is put on production and to a lesser degree on the client and the finished building product. However, in some areas such as heating of the dwelling excellent results have been achieved. For instance the total energy consumption in housing has decreased by 30% since 1972 due to a close cooperation between the public sector, the building sector and houseowners.

There is no doubt that industrialisation within the area of materials production is bound to continue its positive trend without assistance. In the planning and building area, however, there are problems. The development of industrialisation and development of productivity are weak and need more intensive assistance. This is the problem we shall try to elucidate in the following and come up with some suggestions for solutions.

2.1 DEFINITION AND ASSUMPTIONS

When working with productivity and industrialisation together with consultants and workmen these two words often have a negative ring. Mass production in large series of homogeneous products comes to mind or poor and ugly looking products of short durability. Another thought is total industrialisation as being the objective of the building sector and not one of several means of attaining cheaper and better houses.

A common understanding of industrialisation and productivity is lacking in the sector and what meaning we want to attach to it. We will not, in this paper, debate the necessity for such a common understanding, but accept it as a fundamental.

Definition of productivity and industrialisation

In the following we want to consider industrialisation as one of several means to obtain improvements of productivity, and at the same time we wish to work with the final quality of the building, the architecture and aesthetics and the influence on the environment. That means that we are here working with productivity as a wider term for the consumption of resources in the building process and the result in the form of the size of the building, delivery conditions, standard, quality, architecture and aesthetics as well as environmental conditions.

In this context it is not just a question of moving tasks from the building site to an industrial production of finished building products in a factory, but also of industrialising the work on the building site and the planning attached. Therefore industrialisation as a process, on one hand, covers the rational linear production and, on the other hand, also covers the creative design and development processes which are a prerequisite, as well as various variants in between.

Three important assumptions for the productivity development in Denmark

When one looks at the various initiatives of development which have been carried out in Denmark it is striking that the greatest efforts are made in the field of development of products and building elements based on new ideas. Only a small part of the development is to be found in the implementation of the best of the tested ideas and in the field of development of better methods and work routines. Within the development of methods and work routines it is the generating of ideas which is pronounced focusing on the development of various tools to be applied in the execution. That means that development, improvement and implementation of better methods and work routines, in the present Danish development strategy, have a very low priority.

Fig. 2.2. View of the inner city of Copenhagen. In the foreground buildings under renovation as mentioned in the analysis of 88 renovation cases from 1987 - 1995.

This low priority is probably a consequence of the fact that the development activities, which are the easiest to grasp, are furthered quite subconsciously, whereas the activities that are hard to grasp, and which show few results and demand the cooperation of various groups, sub-consciously are given a lower priority. This goes for improvement of methods and work routines, where learning and training are important factors in connection with the implementation of new developments. At the same time it is also a question, whether we today have come so far that this area has become an important factor in the continued development, and that this low priority also reduces the effect of the development of products and tools.

The first assumption therefore says that improvement and implementation of better methods and work routines are important factors in the improvement of the Danish building sector.

The second assumption is, among other things, based on the French scientist and philosopher René Descartes who in "The Method" from 1637 wrote "I started with the most simple objects which were easy to grasp and then progressed little by little to understand the most complex ones". Although this experience has been applied in research, the leap to development and the learning of new methods and work routines is not that great.

When working with improvement of methods and work routines, one soon realises that the power of example is important and that new methods must be made simple and comprehensible if the implementation is going to succeed. An important factor is also the freedom in the choice of method. He who is to work daily with the method should feel that this is the best method for him or for her in the given situation. A better method has therefore not been found until the individual user in a natural way chooses the new method without pressure from the outside.

In conclusion the second assumption says that a new and better method has not been found until it is simple and easily comprehensible and can be made into an example so that the actual user chooses the new method.

The third assumption says that the success of a method depends on the internal cooperation in a building project as well as on clear relations between "customer need" and requirements to individual performances.

Chosen efforts in the support of the development of quality and productivity

In the research undertaken at the Danish Building Research Institute (SBI) we have attached great importance to the study of practical examples which have shown great improvements and which have or might have an influence on the sector as a whole in connection with new building, operation, maintenance and renewal of different types of buildings.

As far as this research is concerned, it has been necessary to develop a common methodology which can be used for documentation of quality and productivity of the different types of building projects. The objective of the development is to find common methods supported by the necessary tools which can be used in the following situations:

- to document individual solutions and comparisons with alternatives
- to analyse and compare several projects and show the current trends
- to point out important areas which can be improved.

The methods should be simple so that they, with a minimum amount of readjustment, can be used both in technological and social research and development, as well as in the development of the various companies in the sector.

It is these definitions, assumptions and chosen efforts which form the basis of the development and the subsequent general description of the SBI-method of analysis of the quality and productivity development of the building sector.

2.2 MODEL FOR ANALYSIS OF QUALITY AND PRODUCTIVITY DEVELOPMENT

The model has been built up based on experiences from quality control, economy control, production control and similar areas in other industrial sectors. It has then been adapted and simplified to a form suitable for the practice of the Danish building sector. The model is based on the scientific tradition where qualitative methods often in the beginning are applied within new research areas and then later exchanged with quantitative methods which, step by step, go from the approximations of the first order to more and more complex models.

The driving force of such a development is the wish to be able to give a better and more exact description of the quality and productive development, but realising that its complexity is impossible to grasp and that it changes constantly.

Fig. 2.3. The street facade of the building in Odense where the method of documentation has been tested.

When comparing such different areas as economy control, production control, quality control and environmental control it turned out that they are all founded on the same general elements. The only difference is that different parameters are used for the control and that these different areas are at a different innovation level. It is also evident that a great part of the data is mutual and that documentation, control and development have many common features. Consequently, it has been the objective of this model to make a common framework which can be used for more than analysing quality and productivity.

The first item of the model is the relationship between the client and the supplier

The model presupposes that the building is to be delivered as a product from the construction parties to the customer, who may be a building owner or an enduser. In this context the word "building" covers both new building and the operation, maintenance and renewal of an existing building. During the building process, it is a question of a chain of customer-supplier conditions covering deliveries of both materials, workman hours, services and building parts.

In a customer-supplier relationship the customer's demands and wishes as regards the product are not always identical with what the supplier can deliver. Possibly a gap of opinion on quality and price between the two parties may occur and be visible when the building is delivered. In the same way the consultants make demands on the contractor and the contractor on the suppliers when the work is being carried out. That means that the chain of deliveries and subdeliveries are presented with a chain of demands on which the two parties may have different views and which on delivery might cause some discussion.

In the Danish building sector these conditions are not so clear as indicated here, but the model presumes that the relationship can be put into words and simplified in this way. In the following the demands of the customer in relation to the total building are not included, only the conditions attached to the building process are stressed.

The second item of the model is the basic elements in the building process

The individual building processes are described based on an input-output model as indicated in Fig. 2.4. The input consists of working hours spent to carry out the working process and raw materials which are included in the final product, and capital including operating equipment supporting the process. The output consists of the product or services delivered from the process, conditions of delivery and the price of the product - the result of the process.

The individual building process and services can be divided into a number of sub-processes which together contribute to the execution of the total building process. In each sub-process a processing of goods and services takes place and between the sub-processes a stream of information and exchange of materials and services take place. In each sub-process and each exchange of information or materials, development or changes may occur. This division of an actual building process into sub-processes, exchange of materials, information and development can be repeated for all sub-processes and thus cover the entire building.

The third item of the model is the controlling elements of a building process

It is a prerequisite for controlling a building process that some requirements and expectations as to the result of the process are made. Towards the end of the process the accomplished values are measured and a comparison of the two values can be expressed in a deviation which, depending on size and importance, may result in the following decisions: No reaction, a regulation of the process or a modification of the expected values.

This means that the establishment of the size of the deviations as well as the choice of a scale of importance as to the importance of the deviation is necessary if a choice or a decision regarding the carrying through of the process has to be made. Even if the Danish building sector does not follow this model even for important building processes, this model has been chosen as the unequivocal reference according to which the building processes in question can be evaluated.

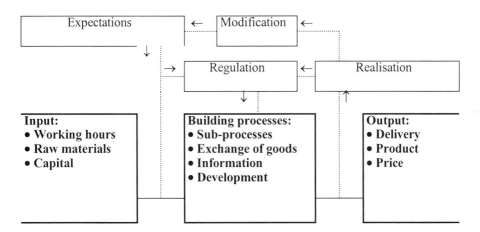

Fig. 2.4. A simple regulation and modification model to regulate a building process which also comprises planning in connection with building.

The fourth item of the model is productivity parameters of the control

The parameters included in the control of a building process must cover the consumption of resources spent in the process and the final result. If these two main parameters are set in proportion to each other you get the parameter model for productivity which in this context describes the building process viewed from the side of the building sector and not from that of the client. In this relation a decreasing consumption of resources will give a rising productivity and an increasing building process result will also give a rise in productivity.

As shown in Fig. 2.5., the consumption of resources is further divided into the three parameters of resources: Time consumption, building products and operating capital. Experiences gained from practical tests have shown that in a productivity analysis it is sufficient to use the three parameters for net prices for time consumption, building products and operating capital, one for the time consumption in a total number of hours and one for the gross price of the total consumption of resources. Based on these five primary parameters, other parameters such as the average hourly wage and the total profit margin can be calculated. A subdivision might be relevant in connection with special studies, for example if further details as to the distribution of the hours spent on mounting, handling of goods, exchange of information and development are desired. In connection with such studies it is recommended that these detailed parameters do not replace, but supplement the primary parameters.

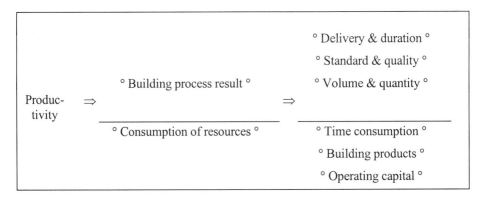

Fig. 2.5. General model of documentation of productivity. The last six primary parameters can all be divided into measurable units.

The numerator of the parameter model which describes the result of the building process is also divided into three parameters: Date of delivery and duration, standard & quality as well as volume and quantity. Date of delivery and duration describes when the result of the actual building process will be delivered and how long it will take to carry out the total process. If, for instance, you wish to carry out more detailed studies, it is possible to make an extension by means of detailed parameters describing the time context between the various building processes.

Under standard & quality the functions and properties as well as architectural and aesthetic qualities are described which are included in the result of the building process. This parameter is probably the most difficult one to handle especially when we talk about the architectural and aesthetic conditions. It is important, however, that some kind of evaluation of these conditions takes place, as a variation of these proper- ties might have a great influence on the consumption of resources. If, for example, one wants to carry out further studies of standard & quality it is possible to make an extension by means of detailed parameters describing the proper building process and not only the result. In the same way it is possible by means of detailed parameters to extend the description of standard and quality to cover the environmental factors and general in- fluence of the process' on the environment. Standard & quality is definitely the least tested parameter and the one which should be deve- loped the most in the future.

The last of the three result parameters - volume and quantity - describes how extensive the work is. How many m^2 floorage, how many m^2 roof have to be manufactured, and how many windows have to be renewed are examples of the description of volume and quantity. When it is a question of renovation, a grouping into types of renovation may be included in this description, for instance, when it is a question of total renovation, execution of minor repairs or if no work is to be carried out. If you wish to carry out further studies of volume and quantity a supplement can be made with detailed parameters describing the procedure and how the work is to be carried out.

It is recommended that the suggested detailed parameters supplement the primary parameters and do not replace them so that a productivity analysis always covers all the primary parameters. Supplementary studies, however, can be made which elucidate special conditions and where special detailed parameters are used.

2.3 PROCEDURE IN CONNECTION WITH THE CARRYING OUT OF THE ANALYSIS

As mentioned in the previous section, the model presupposes that a building can be divided into a number of coherent building processes which again can be divided into a number of basic elements and controlling elements as shown in Fig. 2.4. In the model some common productivity parameters have been laid down, which - cf. Fig. 2.5. - can describe the individual building processes and the building project in its entirety. These two items are further described in the subsequent description including how in practice product and process key figures for the individual building processes are fixed and are to be used for instance for benchmarking. Finally a common model to minimize the deviations is described which can be used when analysing and developing individual building processes.

Division of the project into building parts and contract parts

In Denmark it has for many years been the practice that the various building projects are divided into contracts and often the activities of the contracts have been subdivided according to SfB-classification (CIB Report no. 40, Dec. 1977). In recent years this practice has undergone some changes as several consultants today also use a division of the building project into elements as a base.

One reason might be that hereby considerations are shown towards large builders and end users when agreement in design has to be made and when delivery takes place. This is in contrast to the division in contracts which puts the building process for each contracts in the center. The appearance of improved EDT-systems has supported this development so today it is easy to re-edit the description so that it covers both viewpoints.

Experiences gained from new building and renovation have shown that it is advantageous first to divide the building project into building elements. The division into building elements is the general and stable division focusing on the general agreement and delivery in close connection with the builder and the end user, whereas the division into contract parts focuses on the cooperation between the consultant and the executing party of the building process. That means that the two divisions also are integrated units, each with their own purpose.

Table 2.1. Suggested division of newbuild housing project into Parts and Elements as a basis for specification in productivity documentation.

Break-down of building project	Supplementary description of the individual items
1. Property and land	Total for property and land excl. building
1.1 Building site	Purchase of site, ripe for development, undeveloped
1.2 Connection fees	Fees for connection of installations
1.3 Installations	Installations on the site outside the building
1.4 Ground and planting	Regulation, dressing and planting of ground
1.5 Small buildings	Garages, sheds and similar small buildings
1.6 Property & land and other	Conditions not covered by the above
2. Building	Closed carcass and the other building, excl. rooms
2.1 Foundation & ground deck	All structures and basements "not living quarters"
2.2 Walls	All outer and inner walls
2.3 Deck and horizontal div.	Horizontal divisions incl. decks above basements
2.4 Roof constructions	Roof, attics, eaves, downpipes and facing
2.5 Windows, doors and gates	All windows and openings in outer walls
2.6 Balconies and stairs	Lightshafts, ramps and lifts excl. downpipes
2.7 Installations and plumbing	All installations for distribution and supply
2.8 Building and other	Conditions not covered by the above
3. Flats and rooms	The inner part of the building, excl. walls and deck
3.1 Kitchen	Kitchen with inventory and installations
3.2 Bathroom and lavatory	Bathroom and lavatory with sanitation and installat.
3.3 Other rooms	Other rooms, e.g. living rooms and corridors
3.4 Common rooms	Common room, for dwelling and trade purposes
3.5 Trade units	Premises and rooms in buildings for trade
3.6 Stairs and staircases	Stairs and staircases outside dwellings and trade
3.7 Dwelling & rooms and oth.	Conditions not covered by the above
4. General activities	Common for the project not covered by items 1, 2, 3
4.1 Site operations	Elec. on site, water, sheds, fences and scaffolding
4.2 Design until start	Design and planning until start of construction
4.3 Construction control	Planning and controlling during construction
4.4 Building administration	Administrat. and legal management of building project
4.5 Insurances	Insurances and risk cover of building project
4.6 Financing and duties	Interest, commission, stamp and other financing
4.7 Other activities	Conditions not covered by the above
Total building project	The total building project as a total product

A division into building elements takes place in units which are visible to the building and the end user and have a clear function. The unit shall be clearly limited geometrically, it shall be possible to locate in the final building or in the project, and from project to project the units must be the same for different applications and types of buildings. If, for instance, we take a building project of a new building, a division, as indicated in Table 2.1, is based on the following main areas:

- *Property and land* of the building as well as surrounding areas and installations and small buildings outside the housing property
- *Building*, its outer structures, the load-carrying structures and installation elements for distribution and supply in the building
- *Flats and rooms* of the building such as living-rooms, stairs and common rooms complete with installations
- *General activities* regarding the building project which do not belong to the other items, but concern design, administration and financing of the project.

At the moment a common division into building parts is demonstrated on renovation of small building properties following the principle shown in Table 2.1. It is our intention that a similar division regarding building elements for other important types of buildings will eventually take place both with regard to new building, operation, maintenance and renovation.

Experience tells us that this breakdown is sufficient for a normal documentation of productivity. If it is the wish to obtain more details in special development experiments this can be done by using the structure of the SfB-classification (CIB Report no. 40, Dec. 1977) as a basis for a separation of the individual building elements from the parts of the contract. This would occur in connection with the renovation of small building properties where one objective is a total description and specification of the individual types of installations as a supplement to the indicated breakup in building elements.

Product- and process key figures for benchmarking

With the above we have now laid the foundation of a central task for the reseach of the Danish Building Research Institute (SBI) within quality and productivity development, i.e. to render it visible. In its capacity as a sector research institute it is the duty of the SBI to assist the building and construction sector and society in making better buildings and to do so it is necessary to realise the strong and the weak points of the sector. A means to make the sector visible is to work out various key figures which the sector can refer to - also called benchmarking.

The key figures show both the strong side and the weak side of the capacity of the sector and as these figures change constantly they will be able to pull the strong side and push the weak side along. In Denmark there is a possibility of building up a trustworthy key figure system as we are used to present the state-owned and subsidised buildings as good examples and references.

The key figure system is based upon the fixed quality and productivity parameters with reference to the individual building parts and in relation to the different uses and purposes of the buildings. There are two types of key figures - product-related key figures and process-related key figures. The product-related key figures relate to builder and end user and their communication with the consultant attached to the design and delivery of the building, whereas the process-related key figures relate to the construction parties to further the improvement of quality and pro-ductivity of the individual building and contract parts.

Table 2.2. Product- and process-related key figures for building elements which have been totally renewed at a max level (Analyses of 88 renovated building properties in Copenhagen). Values are expenses on direct construction work in DKK-kroner excl. VAT and in 1995-prices. General activities will add 43.6% to these expenses.

Building elements	Product-related key figures		Process-related key figures	
Exterior building elements:				
Roof	665	kr/m² living area	3.190	kr/m² roof area
Facade	627	kr/m² living area	672	kr/m² floor area
Windows	558	kr/m² living area	558	kr/m² living area
Interior building elements:				
Living-rooms	2.081	kr/m² living area	2.081	kr/m² living area
Bath and toilet	610	kr/m² living area	52.198	kr/ bath-room
Kitchen	458	kr/m² living area	40.050	kr/ kitchen
Installations:				
Heating	423	kr/m² living area	423	kr/m² living area
Water	205	kr/m² living area	17.439	kr/ flat
Drain	181	kr/m² living area	15.489	kr/ flat
Electricity	328	kr/m² living area	328	kr/m² living area

As product-related key figures the net living area or net floor area might be used as a common reference as builder and end user estimate the price and values of the building in relation hereto. The product-related key figures are laid down for the whole building and for the individual building elements, but it is not necessary to go as far as to the individual contract parts. Process-related key figures, on the contrary, are fixed individually for important building elements, construction and contract parts. As an example, the process-related key figure for the entire roof construction can be expressed in proportion to the size of the roof area or the first floor area, whereas the product-related key figure is expressed in proportion to the total floor area. That means that for a one-storey house these two figures will be the same, but the higher the number of floors of a building, the lower the product-related key figure will be.

The key figure system has been applied in an investigation of 88 building properties which during the period 1987 - 1995 underwent a total renovation in different ways. The 88 renovation projects are all divided into 18 different building elements and for each of them a product-related and a process-related key figure have been calculated which are in proportion to the different types and the extent of the renovation work. The most striking type of renovation for the individual building elements is an extensive renewal of high standard. In Table 2.2 a selection of the two sets of key figures for important building elements for such a type of renovation is shown.

The intention is to continue with more investigations to determine how such a key figure system should in practice be set up, and it will be demonstrated in stateowned and subsidised building projects. In this development it is the goal to select some constant key figures and have them tested prior to the implementation so that they remain stable over a long period and can be used by the entire sector.

Minimizing deviations in analysis and development

The key figures are mean figures of examples of similar types of buildings and construction work. In order to render the results of the analysis and the development even more visible it is suggested that also differences and deviations between the individual types of buildings are set down. This can be done by means of an SBI-method called "Stepwise Minimizing of Deviations" (SMOD), a method that has been applied in a number of test constructions in Denmark.

The principle is well-known within quality control and production control, but also within various sports, e.g. shooting. Through an analysis of a number of measurements, the average and standard deviation are to be calculated and put in proportion to the goal. Then a search for different factors which can reduce the standard deviation and the distance to the desired goal has to start. If such an important factor is found it has to be eliminated and one can start a new round of improvements based on the improved results.

The stepwise minimizing of deviations has been applied both in analyses of the total development of productivity of several projects and when analysing an individual project as well as giving priority to future efforts. As a basis for an illustration of the method the process-related key figures for the roof construction can be seen in Table 2.3.

The analysis of the key figures of the roof construction shows a standard deviation of 57 % for the 88 projects in total. By removing atypically expensive projects and projects where no renovation has been done one comes to 82 projects with a standard deviation of 51 %. In this project all the 4 atypical projects have a deviation of between 2 and 3½ standard deviations from the mean figures and they distinguish themselves very clearly from the other projects.

Table 2.3. Stepwise minimizing of deviations (SMOD) in the analysis on roof constructions (Analysis of 88 renovation projects on building properties in Copenhagen during the period 1987 - 1995). The values are expenses on direct construction work in DKK-kroner excl. VAT and in 1995-prices. General activities will add 44 % to these expenses.

Type of renovation and extent	Number of projects	Process-related key figures for roof Expenditure per roof area in kr/m^2		
		Average	Standard deviation	
1. Minor repairs	7	256	55	21%
2. Repairs, min.	2	560	-	-
3. Repairs, max.	2	1.455	-	-
4. Renewal, min.	18	1.755	575	33%
5. Renewal, max.	53	3.190	944	30%
5.1 Non-utilized attic	27	2.759	704	26%
5.2 Utilized attic	26	3.637	964	27%
Sub-total for 82 projects	82	2.518	1.278	51%
Projects not included:				
Atypically expensive projects	4	from 5.187	to 6.946	
Projects without renewal	2	0		
Total for all 88 projects	88	2.615	1.485	57%

The second step is to evaluate the extent and type of renovation and here one arrives at a division into the following 5 types of renovation: Minor repairs, repairs-minimal, repairs-maximum, renewal-minimal and renewal-maximum. By this division the standard deviation for each group is down to 21 - 33%, but the last group - renewal-maximal is very large and consists of a total of 53 projects or 61% of the remaining 82 projects and with a total standard deviation of 30%.

The third step involves finding a way to divide the 53 projects into smaller groups to reduce the standard deviation. One looked for differences between the properties such as for instance number of floors, the age of the building, the size of the building and the utilization degree of the top storeys, and it turned out that the last parameter had a marked influence. The reason is probably extra expenses for attics and sky lights as well as insulation of the roof structure when parts of the top floor is utilized. It also turned out that if the top floor is utilized the degree of utilization does not vary very much.

The group comprising the 53 was therefore divided into two sub-groups - one for buildings with non-utilized top floor and one for buildings with utilized top floor. The standard deviation for the two new and even groups was reduced from 30% to 26% and 27%. That means that now 6 different types of renovation have been found where the standard deviation is down to 21 - 33% for all groups. Simultaneously it can be seen that it is primarily a question of 3 types of renovation, which are worthwhile looking into in the future development of productivity, and that is the 3 groups for renewal (81% of all projects).

2.4 APPLICATION OF THE METHOD FOR DOCUMENTATION OF A SINGLE PROJECT

The method has also been applied to a 4-storey building property with 9 flats in Odense, which were totally renovated in 1995. In the test it was the wish to examine the quality of the design process by looking at the modifications which occurred from contract with a contractor until the delivery of the building. In the investigation it was important to find the areas where it would be advantageous to start improvements or initiate further investigations.

Table 2.4. Realised workman hours distribution on the individual building elements and contracts in the test project, Godthåbsgade 61A, Odense. In the areas marked with an '*', no work was carried out. All values are in hours.

Building elements	Contracts in the renewal projects					Contract total
	Mason	Carpenter	Painter	Plumber	Electrican	
1. Roof	324	966	*	302	*	1.592
2. Facades	848	170	74	*	*	1.092
3. Stairs	47	280	65	*	24	416
4. Bath and toilet	303	73	207	333	28	944
5. Kitchen	139	750	153	10	67	1.119
6. Heating	*	*	16	587	*	603
7. Living-rooms	*	110	408	86	174	778
8. Building site	*	*	*	*	13	13
Elements total	1.661	2.349	923	1.318	306	6.557

Relative deviations in the planned workman hours

The project was divided into 8 building elements and 5 contracts and into many different activities within each of these. Out of the 40 combinations of building elements and contracts, work was only carried out in 27, and for each of these, expected and realised data were collected for each productivity parameter, cf. the model in Fig. 2.5. On this basis the relative deviation for each combination and parameter has been calculated by means of the following expression:

Relative deviation = (Realised - Expected) / Expected

In Table 2.4 an example is given of how the result can be presented for the realised workman hours, and Table 2.5 shows the calculated relative deviations in the same presentation. Table 2.5 is used for pointing out the areas where improvement efforts probably would be of most value. This is done in Table 2.5 by finding the greatest deviations for the combinations of a considerable size. In Table 2.5, 5 areas have been pointed out, which are marked in bold, and these cover a total of 37 % of the total number of workman hours.

That means that if it is possible to find the reason and limit the deviation in these 5 areas, a marked improvement of the result of the design and planning will take place as far as future projects are concerned.

In the same way all the other parameters have been studied one by one to find possibilities of improvement and the same presentation has been used to make it easier for the parties to find the most important areas of future efforts.

Table 2.5. Relative deviation (Realised - Expected) / Expected)) for workman hours in the test project Godthåbsgade 61A in Odense. In the areas marked with an '*' no work has been carried out. The deviations written in bold are the most important ones (37% of all hours), which should be submitted to improvement.

Building elements	Contracts in the renewal projects					Contract total
	Mason	Carpenter	Painter	Plumber	Electrician	
1. Roof	-5%	**119%**	*	-27%	*	33%
2. Facades	-10%	10%	1380%	*	*	-1%
3. Stairs	18%	**93%**	-33%	*	-20%	33%
4. Bath and toilet	12%	-61%	-10%	-47%	-70%	-33%
5. Kitchen	-1%	20%	76%	0%	-28%	17%
6. Heating	*	*	-20%	**90%**	*	83%
7. Living-room	*	-37%	**308%**	115%	**335%**	120%
8. Building site	*	*	*	*	160%	160%
Elements total	-4%	-36%	71%	-6%	17%	16%

Potential cost reductions concerning wages compared to building products

If, on the contrary, one has to compare one productivity parameter with another to find out where future efforts are best applied, a comparison of various divisions can be made by means of the same data. As an example, the difference is shown between the labour cost and the expenses for building products.

In the renovation project in Odense the labour cost amount to 47%, the expenses for building products amount to 49% and operation cost amounts to 4% of a total realised workman expense of 2.690.670 DKK-kroner. That means that improvement efforts regarding the consumption of resources should either focus on labour cost or on building product expenses. If one looks at the relative deviations for both labour cost and building product expenses it is possible to set up two distributions which can be compared, Fig. 2.6 and Fig. 2.7. In each distribution the deviations for the individual building elements, contracts and activities are weighted in proportion to their share of the total expenditure.

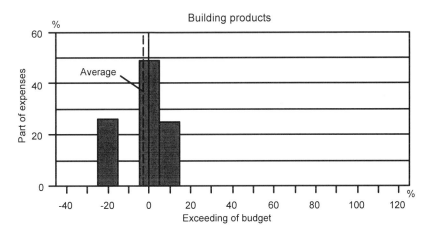

Fig. 2.6. Distribution of budget-exceeding for building products, which amounts to 49% of the total workman-expenses.

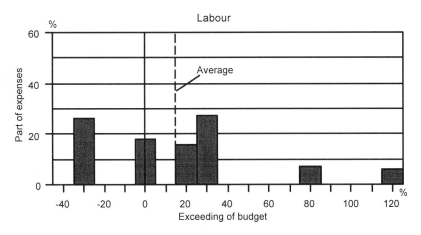

Fig. 2.7. Distribution of budget-exceeding for payroll expenditure for labour, which a-mounts to 47% of the total workman-expenses.

As far as building products are concerned the average is seen to remain at a lower consumption of 3%. For 49% of the expenses for building products the expected values have hit the realised values, and the largest deviation is -20%. Conversely, the average of the labour cost seems to be in excess of 15%, and the deviation is large. For instance 5% of the expenses have an excess of up to 120%. When comparing the two distributions it is evident that the labour costs are the most difficult ones to plan, and that it will be here that the possibility of finding ways of improvement will be the greatest. The interesting aspect of this conclusion is that it is in clear contrast to the efforts spent on development, the largest efforts normally being spent on the field of products. Perhaps because there are fewer variables and more control possible with production of items than human nature. People are much happier to change products than their own behaviour.

As shown in the above example, the important effort areas are pointing at areas where the deviations are the largest. The procedure of the improvement work and of better analyses is to try to minimize deviations. One can question the accurancy of the scientific approach of the method, but as long as so large deviations exist in the construction sector, the method has sufficient qualities to cope with most of the projects.

3

Construction as a manufacturing process

Colin Gray
Dr. (United Kingdom)

3.1 INTRODUCTION

Construction must be like every other industry and grow by providing even better value through stringent analysis of its costs. Saving a further 30% from current cost levels, ie. Summer 1996, is probably unrealistic. However, the UK industry must address the fundamental problem that it wastes two thirds of the labour inputs in a project. Even this is not quite true. What the UK actually does is pay for a design philosophy based on the provision of precisely engineered buildings which have a wealth of detail, that takes mere detail into the realms of exquisite architecture. Every building aims for this goal. The consequence is that every building is extremely complex to produce and can only be afforded if the input costs are low. By international standards the UK's input costs are a third those of an equivalent developed nation.

The existing techniques have so far only achieved approximately 20% cost benefits, even when applied in their purest and most committed form. A new approach will be required. Revolutionary approaches will be too difficult to achieve. An approach based on extending the best of current practice into that of the best in the world has been considered from observation of manufacturing industry. The UK construction industry has one major advantage, its approach to design. This is world class in the field of building design, both in architecture and engineering. There is often bitter division between the designers and producers because the only way to constrain cost is to use all manner of aggressive procurement and contracting policies, such as; lowest bid, Dutch auctioning and late payment. In terms of the current UK industry this has had a destructive effect because it has accelerated the fragmentation, reduced investment and impoverished contractors and starved specialist contractors of investment funds through short term financial manipulation. Sir Michael Latham (1994) addressed some of the issues in his recent study, but not as a value chain which integrates the design and construction process as one to achieve better value.

Cost reduction is not a new phenominum. It has been a constant pressure since before Dr Johnson said that; *'to build is to be robbed'*. In more recent times there have been a stream of reports which have examined and questioned the cost levels of UK construction. The other problem is that there is not a common data base of project cost. There are no published benchmarks upon which a reasoned judgment of cost can be made. Every project starts from a cost of the proposed design whereupon it is left to the client to assess whether they can afford the project. The industry is not cost led. It does not give its clients a common base from which they know that their project does not deviate. Clients are, therefore, in an uncertain environment and have had to live on their wits to beat the system. Traditionally they have relied upon competitive bidding to keep costs under control, but this is a weak method if the two elements, i.e. the product and the process, are not in the competitive equation. Until recently only the products have been subject to competition.

This paper attempts to bring together the previous studies into an analysis of the available methods to achieving the goal of a 30% real cost reduction. It is a selected review which makes the basic assumption that the quality of the product is a constant.

3.2 STEPS TOWARDS A REDUCTION OF REAL COSTS

Historically cost reduction has been achieved through increasing productivity. The UK has been consistently less successful in this than other countries. Whilst there has been and is currently an increase in productivity it is slower than in other comparable countries and starting from a lower base. Productivity improvement is usually accelerated in times of boom as scarce resources have to be made more efficient, but it is noticeable that in the current recession productivity has been increasing. In a study reported in Construction Manager business analyst Plimsoll found that productivity had increased by £2000 per man on average in 1994/95 (Construction Manager, 1995). The conclusion was that fewer people were left in the industry and that they were working harder. Even so long term productivity improvements are only contributing 2%-3% cost improvement per annum.

Because the rate of improvement in productivity has been so low for many years the UK has one of the lowest wage levels in the developed construction industries in the world. This is not utilised to produce low cost construction (The Secteur Study, 1994). Generally the UK's input costs (at purchasing power parity) are a third of other developed countries and yet the out turn costs are the same or higher. Any competitive advantage has been absorbed. However, when UK workers are working they as are effective as any other worker, it is just that they are not working for very long due to task complexity and organisational failures (Horner et al, 1989).

3.3 PROJECT SPECIFIC TECHNIQUES

There are many improvement techniques available. They are well known by people in the industry and yet they are not delivering the expected savings. Over 60 cost saving techniques can be listed. Each of these have been assessed as to their contribution to saving cost. Generally they individually can contribute between 1%-5% and perhaps upto 10% savings in cost. However, the impact of these techniques is not cumulative and there is doubt that on a conventionally organised project they could collectively contribute more than 10% (CII, 1990).

3.4 CRITICAL SUCCESS FACTORS

Clearly the construction industries in other countries are able to deliver a much higher level of production than in the UK. What are the keys to this apparent success? In a study of American and Japanese construction industries Mathews investigated international competitive advantage and identified several critical success factors plus the two overarching requirements of training and research and development (Mathews, 1994). But, he concluded, these techniques at best contribute 5%-10%.

3.5 CONSTRUCTION MANAGEMENT

In the 1980s some clients' teams borrowed many ideas from the US and the project teams focused on creating an integrated design and construction process which could respond to the vagaries of the UK development market. They sought the best quality designers, managers and specialist contractors. Where these did not exist they were either imported, developed from the existing UK base or created from combining suitable inputs. Every aspect of conventional contracting was questioned and redesigned. The 'can do' philosophy was uppermost in everyone's mind during the whole of the project.. Considerable help was given to the specialist contractors through section induction course into the new way of working and the associated management practices that would be essential to make them work (Bennett and Gray). This approach achieved a 15% cost reduction. It has since been improved to 20% (Gray 1996).

3.6 VALUE CHAINS

What CM has shown is that to be efficient each step in a chain of activity must be linked so that the value added in each step is an incremental improvement upon the preceding step. If the steps are not linked in this way then only the benefits within the step can be obtained. In comparison with other countries the UK construction industry does not achieve the potential added value due to the significant cultural polarisation within the industry and weak linkages between the steps in the value chain. From the productivity point of view the following are the ideal attributes of the consequences of the design and supply chain:

- Repetition - giving familiarity, learning and consistent application of skills
- No breaks - leading to continuity, care of the work, pride in the work
- No delays - giving continuity, no demotivation, no cyclical start delays
- Material supply - no return visits, no disruption, no delays
- Material packaged for installation - no multiple handling, no damage, no delay
- Damage free materials - no reordering delay, no disruption to work flow
- Team working - mutual support, flexible skills, multiple skills
- No interference - leading to continuity, high morale, pride in the work

An effective value chain in construction recognises the inherent link between the design of the components within each technology and the impact on the above.

3.7 FAILURE IN THE VALUE CHAIN

Each participant in the value chain has a relationship or linkage (direct or indirect) with one or more participants which are either: physical, contractual, design or cultural. The opportunities to enhance value creation and comparative advantage within the UK development and construction industry are impeded by imperfections in the value system. One of the primary features of the development of the construction industry is its fragmentation. The fragmentation is due to:

- Physical: participants and resources are freely sourced from around the world as there are limited barriers to entry to the UK.
- By industry: many different industries through their various manufacturing interests contribute to the design and construction of a single project.
- By education and training: there are a wide range of educational and training backgrounds from professional to craft.
- In time: a projects time span often covers 3-4 years or more and different participants are used at different times and often for much shorter periods within the overall time frame.
- By culture: many participants are from culturally dissimilar backgrounds. Culture is a product of many inter-related factors, which form a web representing the culture of an organisation. An organisation's culture can be a very powerful force which generates distinct attitudes and operational practices used to operate and control the individual processes within the chain.

In practice these differences are very real when applied to the chain intending to achieve the objectives set out above:

Step in the chain	Desirable requirement	Barriers
Supply to site	Site led demand, delivery of complete sets for assembly	Minimal stocks, separate transport organisation/ schedules
Manufacturing	Optimise production system, repetition, bulk supply	Customisation, site led demand
	Optimise manufacturing	Sub contracted detailing
s	Optimise site assembly	Sub contracted based on price
Procurement	Integrated working	Fragmentation, lowest price
	Best price	Product/process prescribed
Design:		
Concept	Optimise return/value	Conflict in objectives, no
Scheme	Optimise customer need	link to production needs
Detail	Optimise material	

Each link in the chain has a different set of optimisation goals. Each link in the chain uses different organisations and inevitably the goals change at the interface. Each organisation has its own culture and therefore the whole chain has a cultural dissonance. The overall customer goals set at concept and intended to be executed at construction are lost in sub part optimisation.

"The truth is that **by optimising independently each phase** of the traditional building process, marketing, programming, design, engineering and scheduling and site execution - **one runs the risk of ultimately losing the partial productivity gain** thus obtained."

(Lahdenpera, 1996)

3.8 VALUE TRANSMISSION

The flow of work from one sector to another must be smoothly and efficiently transmitted with no delays or imperfections. Within each cultural group the transfer of value is generally acceptable because communication is good and mutual goals exist. However, between each group transfer is not efficient and this is one of the main impediments to the optimal operation of the value system. Value transfer can be impeded in a number of ways:

Technological continuity
The fragmentation of the industry by geographic separation, sub-contracting within the group, time of involvement and culture creates

many hurdles over which ideas, design concepts, detail designs and thus value creation have to be passed.

Cultural dissonance

For efficient transfer of value to take place, the output from one participant must correlate well with the input of the downstream participant. Differences in product and system knowledge and lack of common goal will prevent accurate correlation.

Organisational continuity

The existing management systems are weak at creating an integral whole in which the value chain is not fragmented.

3.9 TECHNOLOGY CLUSTERS TO ACHIEVE VALUE

A strategy which emphasises cost to the exclusion of quality is not one which is acceptable in the UK. Quality in this context is the attention to detail that produces a crafted product with manufactured components. In this the UK excels, but the consequence is a most complex production task. The future, therefore, must be tailored to the development of a management capability which can deal with the implicit complexity of projects.

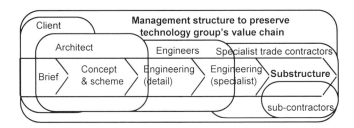

A technology cluster addresses a complete set of functional systems within the building involving one or an interrelated set of disciplines. Within each technology cluster the client, designers, managers and specialist trade contractors are involved from day one in working closely together to provide the best solution to meet the client's need. The technology cluster maintains the value chain of the whole design and production process and this is the first of the essential keys to success over the existing systems - there is no break in the value chain.

A building may have five or six technology clusters which must be coordinated as a whole to ensure that the individual value chains are not broken. A tripartite group of the client, lead designer and manager must provide this overall coordination at every moment in the project. They ensure that the overall goals predominate all activity and the total value chain is not broken. This is the second essential key to success.

Each project must set out to achieve big targets or goals. There is no time on the average project for long term learning. This is the responsibility of the supporting organisations. The quality of the supporting organisations is the third key to success. As the 30% target is so large, given the industry's capability over the last 40 years, it is likely that only a few organisations currently have the capability to support this strategy. It will be totally dependent on the development of super clients, super designers, super managers/contractors and super trade contractors.

3.10 COMPONENT BASED TECHNOLOGY

In the UK the design team is the natural innovator. As shown by the number of drawings in the sample buildings the design team is taking the lead in all decisions affecting the project. The architects see themselves as the lead innovator and the other members of the team see them also in that role. This is relatively unconstrained as meeting the client's requirements are the primary goal.

Technology is used in an uncompromising way to achieve the satisfaction of the goal. It has to be immensely flexible as does the system which provides it. Designer's need access to the manufacturers so that they can experiment with the technology to achieve the best application. This is a high risk for both the designer and the supplier as they are both working in an uncertain environment until the component is installed in the building and integrated with the other components in the final assembly. At the moment the risks are controlled by either the designer proscribing all aspects of the technology in which case the designer must be technically literate and competent or the designer works in close harmony with the specialist supplier to optimise the production and installation process. The implication is that designers must become technically literate in a vast variety of technologies to a much higher extent than they are now, or the component supply industries must provide the data in an interactive way to allow the designers to make the right decisions when proscribing the technologies.

Manufacturing systems

Ideally a manufacturer would like to optimise the factory by producing identical products on a continuous basis. This gives economies of scale and thus economic prices. It also enables continual improvement through process analysis and investment in new technology and skills to achieve development of the products. Where this approach is fragmented with more and greater variety of products from the same manufacturing process then investment, improvement and development becomes more difficult and expensive. Also the process becomes much more difficult to manage.

Customisation

Customisation, at the moment is a high cost option. It affects both the materials cost as the majority of components are one-off and the consequential process is complex and prone to failure. However, the problem will persist and manufacturers will have to seize it as an opportunity to develop approaches to optimise their capabilities, whilst still providing individual solutions.

Standardisation

Some products and components can be standardised and where this is the case their costs must be gradually reduced, in real terms, through greater efficiency. However a wholesale switch to standardised components is unlikely to occur unless there are some major changes in design policy and these will only occur if there is a major constraint on cost or construction resource in a period of boom.

Standardisation requires huge volumes to justify the setting up of the production facility together with a stable long term demand. Stability can be provided by: limiting choice, but the UK has so much choice anyway that it is unlikely to give it up; or increasing demand, which is closely allied to the overall economy and is, therefore, probably not a viable strategy. Designers can help by not fragmenting the demand with specials, but this will be against the trend.

Tight/loose

The way of optimisation between standardisation and customisation is to work tight/loose. The initial design and specification must allow the subsequent specialist expertise to optimise their own skills to achieve efficient and economic manufacturing. The skills and knowledge necessary to develop a component must be defined as should the source of the expertise. The interface boundary is tightly drawn and is the task of the design team, client and manager. By defining the interfaces and not stepping over them into the detail of the subsequent process the initiator of the specification allows the manufacturing organisation to utilise its skills to the best advantage of the company and the project.

3.11 WORLD CLASS SPECIALIST TRADE CONTRACTORS

Continually enhance technology

Whilst designers are the primary innovators in the UK construction industry they need to be fed and supported by an ever increasingly rich set of technologies. The STCs have the responsibility for developing the technologies. They must embrace the working within the technology clusters as the only way in which, on the one hand they can access and learn from the development of their technology and on the other hand resist some of the pressures to produce a totally customised solution which in the end reduces their ability to offer economic products. This dichotomy between standardisation and customisation will be a constant battleground unless all sides of the argument are brought to the discussion. Value is currently lost because the discussion is one sided.

Pick the best people

STCs must have the best people. They are the organisations that provide the productive capability to the project. The scope of their work is constantly increasing as more is passed onto them. As project and consequently sub task complexity increases the STCs must pick the best of their current people and include them in the technology clusters. However the industry is very weak in this area and there must be a concerted effort to raise the profile of this group of people to the intellectual status of the rest of the project team.

Train constantly

As the technological core of the project is with the STCs then they must train continuously to enhance their own technological skills. Their people must be able to integrate project learning into a corporate learning and so enhance the service they give to other projects. The organisation itself must have a corporate structure which captures and disseminates project learning to its other staff.

Be financially healthy

Super STCs are able to differentiate their product through innovation and by offering a better service than the competition. They then secure premium prices. This has been the strategy employed by many which has lead to the rich variety of components and products available on the UK market. It is a profitable strategy which must be accelerated to provide even better capabilities to support innovative UK design.

3.12 IMPLEMENTATION STRATEGY

The UK has a rich design culture based on innovation, attention to detail, the customisation of design to suit the customer's ultimate need and very low input costs.. The cost reduction scenario based on comparison with the USA would not be acceptable as that is based on a stripped down approach to design using relatively undeveloped technologies. A method of construction must be developed which maximises the capabilities of the UK's world class designers. Three areas must be addressed simultaneously: use the most advanced technology, have an organisation which pushes and supports designers to be the 'best' and create a culture to support their innovation. To do all of this at the purely project level will not work. There are too many opportunities for it to break down.

A system must be created which:

- capitalises on the best of UK design skills,
- manages complexity,
- maintains the value chains,
- integrates the client, designers, managers and specialist trade contractors, and
- fosters technological excellence, focused organisation and innovative attitudes in all.

The proposal is to split the project into technology clusters which achieves the necessary joint focus to obtain the integration necessary to support the designers. The client, designers, managers and specialist trades are represented within each cluster. The manager's role is to orchestrate the production of innovative and cost effective design by fostering the close working between the specialist contractors and the design team. However, current practice does not do this in a way which focuses upon the systems in the building nor achieves the level of integration both physical and psychological that is necessary to resolve the complex issues with sufficient commercial rigour. Isolated examples have shown that where this has been achieved then it is a very effective way of achieving best UK practice.

The challenge is turn this into a general project strategy for all projects.

REFERENCES

CII, (1990) Constructability White Paper, Texas A & M, Austin, Construction Manager, (1995).

Bennett, J and Gray C (1994), Construction Management, in The Storey of Broadgate, Stanhope Developments plc, London, unpublished.

Gray C (1996), Faster, Better Value Construction - a review of construction management practice, The Reading Production Engineering Group, University of Reading, Reading.

Horner M and Talhouni (1995), Effects of accelerated working, delays and disruption on labour productivity, Occasional paper, CIOB, Ascot.

Lahdenpera P, (1995a), Reorganising the building process - the holistic approach, VTT Publication No 258, VTT, Finland, p. 67.

Mathews G (1992), Barriers to value and competitive advantage in the UK property industry, Lynton plc, unpublished.

Latham, Sir M (1993) Constructing The Team, HMSO, London

The Construction Secteur Study, (1994), EU, Brussels

4

Open and industrialised building as a part of the comprehensive development of the industry

Pertti Lahdenperä
D.Tech. (Finland)

4.1 INTRODUCTION

The breakdown of tasks between, and the organisation of, the parties to the construction process correspond to the know-how construction industry firms have, and vice versa. Simplifying considerably, we can say that although there is an abundance of contract forms for the implementation of building projects, to a large extent, different modes are based on a quite standardised division of know-how and labour on the lower levels of the building process and, primarily, only contractual relationships vary.

On the other hand, the prevailing building process is beset by many problems. As a result, each of the parties to the building process has developed the process, more or less, but for his own part only, and only from his own needs and views. In other words, the long value-added chain of construction together with the clear-cut roles has led to partial optimisation while none of the parties has been interested in the comprehensive development of the sector. Thus, in spite of the efforts, the problems have resulted in high costs of construction and buildings as well as in end-product quality that does not match the client's needs.

For that reason, we have to strive for quick remedies within the prevailing structure of the building industry, while preparing also for more radical changes in the long run. In fact, due to the observed problems, and the fact that the prevailing organisation won't perform optimally in its future environment either, or especially in it, it seems as if comprehensive renewal of the organisation of the building process as well as a more or less fundamental restructuring of the building industry are unavoidable.

This was a partial conclusion of the study (Lahdenperä 1995) presented below. The aim of the work was to define the most appropriate generic structure of the overall building process organisation for the western countries in the future, i.e. in the long term. The practical aim of the research was to produce a vision of the division of labour between firms engaged in the future building construction project and their operational models.

This kind of overall organisation of the building process involves the overall optimisation of several divergent partial problems representing different viewpoints with multiple goals. The task is highly demanding in its diversity when starting from theoretical premises. Therefore, it was decided best to start out by surveying the potential solutions presented in various contexts and by evaluating, organising and combining them from the viewpoint of the overall goals of the research. This enabled using much of the development work already done.

In fact, hundreds of construction experts — researchers as well as experts on the practical side — have indirectly contributed to the development of the various proposed solutions, i.e. country-specific philosophies, while the ideas have also been tested in many pilot projects. Thus, the study reflects the results of quite important and wide-ranging development work despite its apparent narrow scope and there is even wide agreement about the philosophy-specific solutions.

The national inventories of the host study (Bakens 1997) and the discussion in the international study group (within the CIB W82) led to the selection of four building production philosophies. The bases from which the philosophies have evolved are quite different and the problems related to them can be quite diverse. They can be summarised as follows when we exclude individual problems and the principles for their solution:

- **Finnish system unit contracting** is the approach which focuses most exclusively on the analysis of the production organisation. The fundamental idea is to improve overall productivity and quality primarily by clarifying liabilities and improving motivation which also improves the preconditions for development work.
- **The French sequential procedure** is based on organisation of work in the implementation phase, though it has expanded into versatile development of production. The fundamental idea is to improve

productivity and cut waste in the process while minimising risks in general and creating the preconditions for effective utilisation of technology.

- **Japanese computer-integrated construction** is obviously highly technology-centred. The fundamental idea is to improve information flow and process control using new technology and to speed up implementation. Other very important goals are better working conditions and minimal dependence on labour.
- **Dutch open building** is mainly concerned with the customer's needs and changes in them, even during the erection. The fundamental idea is to produce efficiently buildings and spaces of prefabricated components that meet the customers' individual needs. The spaces must also be modifiable which again emphasises life-cycle economics.

The paper first gives a general description of the problems related to the traditional process as experienced in the context of the philosophies (chapter 4.2). It also briefly explains the study procedures (chapter 4.3). Chapter 4.4 presents the principles of change derived from the research whereas the general applicability of the results is considered in Chapter 4.5.

4.2 WEAKNESSES OF THE TRADITIONAL PRACTISE

In traditional building the individual needs of various customers occupying various spaces of the building cannot be taken into account well enough, since the plans are fixed already at the initial stage of the project — even as to details — whether the end users are known or not. Changes in plans during construction result in additional costs and disputes. In addition to the construction phase, the changes in the needs of users during the life cycle of a building, and changing users, have been poorly considered in construction as future changes and renovation can be expected to be unnecessarily expensive due to the large amount of heavy construction that needs to be done.

Further, the complicated building process is, primarily, the result of differentiated design and production. Especially if planning is tightly controlled by the owner who gets involved in technical details and applies competitive bidding, co-operation fails, solutions become difficult to implement and cost savings cannot be realised since it is not profitable to the bidder to develop better alternatives that alter the plan, and it is not worthwhile improving services and quality beyond the set level as price is the crucial factor. The end result is undervaluation of development work, service and quality in construction. In other words, competition is limited.

In principle, some of the problems might be eliminated if design were the responsibility of the main contractor. However, in the traditional mode this has created new problems related to the functionality and aesthetics of the building when production dictates how building is done. In both cases the use of general designers leads also to unworkable technical solutions in the case of systems requiring special know-how. Changes during construction and re-designing cause excessive disturbance to production as well as in the form of contractual disputes and other compound effects.

Further, the scopes of liability of various teams and deliveries are traditionally based on professions, and do not conform to buildings' actual functional and productional entities. Thus, they are a barrier to development. It is difficult to get feedback and there is no accumulation of knowledge as know-how is possessed by individuals and goes with them. The uniqueness of construction emphasises these obstacles.

The situation is further complicated by the fact that the various stages and tasks of building are highly interdependent which creates a vicious circle. Thus, any disturbances are widely reflected on the activity of other parties causing compound effects. Combined with the synchronisation problems created by rigid professional divisions of labour and the one-time nature of construction projects, such disturbances are quite likely, which is a reason for the weakened motivation of site workers and super-visors.

4.3 DERIVATION OF THE CHANGES

The bases from which the philosophies have evolved are, as explained above, quite different and the problems related to them can be quite diverse. Thus, the philosophies also show and solve the problems of construction from many sides which is naturally very good from the viewpoint of getting optimal research results. Basically, this kind of purposeful selection is a compromise between the organisational viewpoint, future orientation and economic, cultural and climatic factors, etc. In other words, the organisational approach couldn't be the only criterion because of the multiple goals of the study.

It must also be repeated, concerning problem symptoms as bases of the development of the philosophies, that they cover both the observed problems of current practice as well as expected changes in the future environment likely to constrain the building process. From this point of view, the so-called problem-solution analysis was conducted by production philosophies to study the argumentation chains used for analysing problems and solutions in construction. The decomposition offered also a good framework for the derivation of common principles of change, which was implemented mainly by surveying principal and organisational solutions of the philosophies. In practice, all parallel principles of various philosophies were grouped together. Further, the study surveyed the compatibility of various principles of change by the cross-impact analysis, which seeks to integrate all the trends to find the net solution. Also, principles of change that superficially thinking seem independent have significant connections.

Thus, in addition to the fact that various principles seem to make the construction process more effective as independently implemented solutions, more or less, they also made the implementation of other principles possible for their part. In the final end, we seem to be dealing with a comprehensive, new and practical production philosophy and not just a group of compatible principles for making the process more effective. In addition to the strict methodological basis, the evidence for the correctness of the results was derived from practical experiences in target countries as well as by numerous corresponding partial proposals and experiences from other than the four primary target countries.

4.4 CHANGES IN THE BUILDING PROCESS

The study proved clearly that, in general, the traditional building process organisation doesn't perform optimally in its present environment nor, especially, will it in its future environment. The building process and its organisation have to be changed to reap maximum benefits which requires considering the building process an integral whole. This means that new and better models can only be applied if their introduction is combined with a more or less fundamental restructuring of the building industry.

The presentation of conclusions about future organisation is made in relation to the present prevalent procurement method, traditional design-bid-build contracting, including separate design and competitive bidding based on price, work-specific labour-only-subcontractors implementing the building through intensive on-site work, etc.

Thus, if the criteria are customer orientation and quality, efficiency and productivity, and capacity and preconditions to develop — with their manyfaceted meanings — the following ten change principles are necessary for the organisation of the traditional building process and industry in the western countries:

Change from the client's viewpoint

1 `Consumer-oriented phased and focused decision making ´. The opinion of the customer (and various groups) is incorporated through decision-making at various levels and phases. Thus, decisions about the internal systems, for example, will be individually taken by the actual customer, the immediate user, at a late phase of the process. Alternatives are offered and visualised for the customer in question and each of them can make his choice from a reasonable number of available solutions. The procedure provides solutions that meet the customer's individual needs while the phasing of planning and its overlapping with production enables a shorter project implementation time.

2 `Distinction between the shell and interior of the building ´. Technical differentiation between the shell and interior of the building is necessary primarily to better take into account individuality and to gain needed modifiability. It results in different entities from the viewpoint of production technology. The production technology and know-how for them can best be developed in separate organisations based on clearly defined tasks. Also, since the shell and interior often have different clients, the former's client being a larger entity and the latter's a smaller subgroup or individual, their implementation is differentiated also organisationally and, to some extent, time wise.

3 `Performance approach in planning and specification ´. In order to specify the actual goals of the building process, i.e. the meeting of the needs of the user, and to clarify planning and analysis of alternatives, the independent architect employed by the client defines the building, its spaces and systems, functionally and aesthetically, i.e. independent of material- and technology-bound solutions. This includes measurable functional requirements and selection criteria. He consults the technical experts in questions of technical nature. Besides, the need survey is more deep going and systematic than earlier.

Change from the competition viewpoint

4 `Competition based on implementer's technical solutions´. Contractors compete on the basis of technical implementation solutions that meet the client's performance criteria and are results of co-

operation networks they have created, product development and tender procedures. Thereby, firms are motivated to develop production technology and the product as an entity. The basic factors of competition are the so-called quality-related factors derived from the properties of the end product: performance, quality level, safety and durability, pleasantness and aesthetic values, individuality and image, modifiability and consideration of environmental values.

5   `**Extended commercial means of competition** ´. Competition based on product properties produces various innovations and product solutions. For motivational reasons, it must be possible for the supplier to be able to take advantage of the product's possibilities in competition. On the other hand, convincing the client of the superiority of the product requires also that the supplier is fully committed to his product and its implementation. Thus, construction time, a non-disturbing and safe construction process and guarantee period — even maintenance — can be means of competition besides price. Collateral and financing arrangements combined with procurement of tenants and income from the spaces are also possibilities.

6   `**Activation of research and development** ´. As factors of competition become more versatile, R&D becomes a key strategic area for companies and it also largely replaces project-specific planning. The development of the product concept itself and the related production technology and application systems and directions as well as production management systems will be a significant part of the tasks of the implementer to ensure his and his products' competitiveness. The effort increases also the firm's productivity in a way which is no longer in conflict with the goals of the process as a whole.

Change from the production viewpoint

7   `**Establishment of system units for assigning scopes of liability** ´. Organisational scopes of liability are made to conform to system units instead of tasks, the former being as independent functional and productional entities of the building as possible with variation between systems and hierarchical levels. Thus, one organisation is liable for design, fabrication or procurement, and logistics and assembly, including also most site equipment and auxiliary construction work needed to make their system into a finished product connected to the building. The main implementer, on the one hand, and the architect-led planning team, on the other, ensure the compatibility of various system units.

8 `**Industrial component production** ´. Value-adding building work is largely being transferred from the difficult site to more controlled circumstances in prefabrication plants. After prefabrication of various components, that can largely be produced in parallel, and possible preassembly, the work on site is pure assembly in protected conditions. Automation and robotisation is increasingly being introduced in plants and is becoming a possibility also on buildings sites. The functionality and individuality demanded by the client are achieved through varying combinations and accessories of standard components that are part of the product systems developed by firms.

9 `**System-unit-based multi-skilled teams on site** ´. The tasks of the new types of building teams and trades are organised around the method needed to finish the functional and productional entity, the system unit, instead of having repetitive, monotonous, and specialised works being done by trades without a clear scope of liability with respect to the final product. These new teams expand their job description to include new physical and planning tasks. Thus, more multi-skilled team members are needed so that every team member knows all the tasks of the team, in principle. This eliminates the waste of time and the distinction between planning and control and physical labour and boosts motivation and smooth implementation of the project.

10 `**Continuous collaboration between parties** ´. Continuous collaboration within and between companies increases and standard procedures are developed. On the one hand, this means working as partners from one project to another in many cases, which increases productivity through the learning effect and various innovations. On the other hand, the continuity of co-operation offers a chance for joint development projects where integration of wide-ranging know-how and the various novel viewpoints of different types of organisations are of great benefit, as well as the attendant unbiased attitudes.

4.5 DISCUSSION

The completion of the changes suggested as a result of the research has been based not only on the problems of the present practice but also on different general development trends and phenomena existing in the market and society in general, which supports the idea that the solution as such is a mature vision of the future forms of organising the construction process — and utilising the open and industrialised building principles as part of it.

However, some of the general trends — increasing competition, individuality and speed of change — mean that new products and modes of operation are being developed, tested and introduced. Together these factors increase the diversity of the construction process, especially since different firms choose different means and stress different factors of competition as new possibilities open up.

Also, a typical characteristic of construction in comparison to other industries is its dependence on clients' individual demands, and local conditions and culture, which even strong internationalisation cannot eliminate. Moreover, projects in general, and especially those involving different building types, vary a lot and have different goals and constraints. Thus, the suggested changes cannot make a breakthrough everywhere as such or, especially, as similar in all details. Non-adherence to individual principles might well be practical in some instances. All in all, building production is a quite multifaceted sector and will obviously be even more so in the future.

Despite the diversity, increased organisation of the construction process and corresponding structural change in the construction industry according to the above-described principles of change must and will be the absolute main direction of development generally. This way the huge improvement potential of the building process can be utilised while the changes create an organisational environment which stimulates innovation.

Correspondingly, the wideness of the above definitions leaves many open questions and unsolved practical procedures regarding the details of the solution principles. Thus, methods and tools as well as attitudes and capabilities must be developed in order to reap the available benefits. The abundance of subject matters covers those of general or common interest as well as the operational methods and technology of enterprises.

Various obstacles, such as traditional attitudes of professionals, will also hamper the development. Overcoming the resistance to change may be extremely slow if the superiority and low risk of the operational methods are not convincingly established. This, again, may prove difficult since temporary project-specific — or otherwise seldom used — roles that appear to differ from the traditional ones do not yield the project the benefits that are attainable through longer-term change of procedures. This is a result of the fact that the know-how of traditional companies and professionals does not correspond to the requirements of implementing the new process.

The evolution of `complete´ know-how in organisations based on the new roles will thus take much time and need laborious development. The evolution will not happen overnight. The selected strategies and subsequent procedures of the spearhead companies play a key role in this respect. Subsequently, when these improvements are used for competitive advantage, more and more firms will be forced to implement similar measures — if they hope to be involved in future construction.

REFERENCES

Bakens, W. (ed.) 1997. Future organisation of the building process. Final report. Rotterdam, NL: International Council for Building Research Studies and Documentation (CIB). 270 p. CIB Publication 172. ISBN 90-6363-005-0

Lahdenperä, P. (1995). Reorganizing the building process. The holistic approach. Espoo, FI: Technical Research Centre of Finland (VTT). 210 p. + app. 7 p. VTT Publications 258. ISBN 951-38-4796-9

5

A vision of future building technology

Asko Sarja
Professor, D.Tech. (Finland)

5.1 INTRODUCTION

Building technology, more so than other branches of technology, is strongly tied to the surrounding society (Fig. 1). Future visions of building technology are therefore different in different societies, which fall roughly into early industrialised countries, newly industrialised countries and developing countries. This vision focuses on early industrialised countries, but in the longer term is also valid for newly industrialised countries. In developing countries the building technology will probably take different directions in the long term, although some similarities might exist.

Figure 1 (Sarja, 1989) /1/ describes the general mechanism of development as follows: The objectives and targets of construction are defined by the general politics of each society in relation to the environmental, sociological and economic objectives. Underlying this development and specific to each region and society is a heritage of culture, arts and architecture. The advancement of knowledge is served by natural sciences, engineering sciences and general technology and is transferred to building technology through building design, production and products. The entire evolution is strongly governed by market forces and company operations.

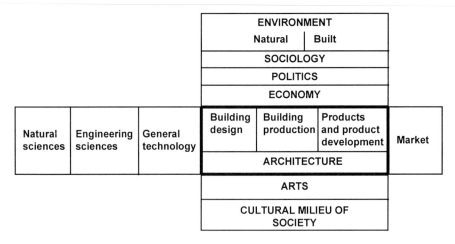

Fig. 1. Development environment of building technology /1,2/.

In developed countries the working environment and the role of construction are undergoing rapid change. Reasons include the passage of societies into a post-industrialised information era and the increasing regional and global economic integration between national economies within the framework of an increasingly global economy. These changes are creating new challenges for the construction industry, which can be met successfully only through active and innovative changes in building design, products, manufacturing methods and management. Introducing into these challenges the latest advancements in technology will yield drastic developments and very likely a totally new generation of building technology within the next 10 to 15 years. New solutions will be developed, or already partially exist in the form of prototypes and limited applications. This evolution, which started in the second half of the 1980s, will gradually penetrate into wider applications until the new generation of building technology is common practice, most likely by 2010-2015.

5.2 GOALS AND THE TRANSITION TOWARDS THEM

The main goal of construction in all societies is a good quality of the built environment in terms of aesthetics, health, economy and ecology throughout its lifespan, fulfilling clients' needs and all the requirements of a sustainable society and nature. In recent years awareness has increased worldwide of the importance of the environment and ecology in the long term development of construction. To meet the goal set out above, the real content and methodology for sustainability must be concretised and introduced into practice through development of design, product systems and products, manufacturing methods and site construction methods.

The general starting point of sustainable building is the definition given by the World Commission on Environment and Development in 1987, whereby sustainable development is "meeting the needs of the present without compromising the ability of future generations to meet their own needs". Sustainability for a society has two components: sustainable physical relations of society with nature and internal sustainability within the society.

The key point in understanding the concrete content of sustainable building is the interpretation of the general principles into construction practice. This means that we must recognise different aspects of sustainability in building and integrate them into a comprehensive set of general requirements and further into more detailed specifications. The physical relationships of construction with nature give rise to requirements pertaining to the ecology, health and the physical environment. Internal sustainability within the society and the building sector give rise to requirements concerning economy, functionality, performance, durability and aesthetics. This is the scheme presented in Figure 2 (Sarja, 1997) /3/. Sustainability must always be treated according to the life cycle principle – in other words with the application of life cycle methodology to design, manufacture, construction and maintenance, as well as in the management of building projects, companies and of other organisations of building partners.

Referring to Figure 2 we could give a technical definition for sustainable building as follows (Sarja, 1997) /3/:

"Sustainable building is a form of building technology and practice which meets the multiple requirements of people and society throughout the life cycle of the building."

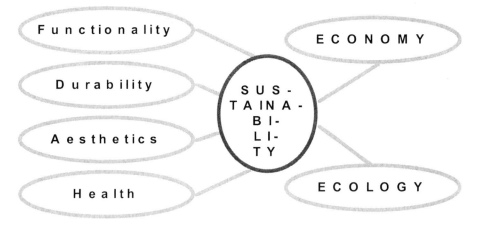

Fig. 2. Multiple requirements for sustainable buildings.

In the new generation of building, architects, technical designers and production managers will work within a newly organised co-operation framework which will help them realise their maximum creative and innovative potential. Both workers and students will feel privileged to enter a prestigious rewarding, healthy, safe and secure career in construction. Construction will be highly industrialised, combining the use of advanced design and production planning methods with mechanised and automated manufacturing and construction methods. Difficult, dangerous and heavy site tasks will give way to factory-produced components and modules and mechanisation of the site process. Craftsmen will be free to use their improved skills productively. The increased investments of factories will be used in a flexible way, allowing rapid changes of production capacity through the increase or decrease of working shifts strongly undergirded by automation. Expansion of market areas across borders will enable more stable markets for construction.

The transition into the new generation of building technology is a long term process, in which product development and the development of design and management methods will be the most important links. These will introduce environmental sustainability principles and systematic methods of analysis, optimisation and decision-making into the practice of all partners of construction, taking into account the entire life cycle and recycling of buildings, technical product systems, components and materials. The real technology is changing through long term development of design methods, building systems, building concepts and building products. Such an evolution will give excellent value to buildings which are both economical and ecological throughout their life cycle. Advanced material, structural, machine, equipment, automation and information technologies will combine to form an integrated whole in advanced building (Sarja, 1995) /2/. A new building infrastructure will be created through organisational, management and business development.

5.3 COMPONENTS OF THE FUTURE BUILDING TECHNOLOGY

As described above, the components which will dictate future sustainable building technology are integrated life cycle design, open and modular production, integrated information and business networking.

5.3.1 Integrated life cycle design

5.3.1.1 Design process

Controlled and rational decision-making when optimising between multiple requirements with different metrics is possible only through application of the systematics of multiple requirements decision-making. Ecology and health aspects are of ever increasing importance and service life principles are introducing time as a variable in economics and design. Close co-operation with clients and architects is imperative. All these aspects are widening the scope of structural design and construction to the extent that entire working processes must be re-engineered. In design, we can start to establish a new design process, or so-called integrated structural design, which is scheduled and described in the following.

Integrated structural life cycle design includes the following main phases of the design process (Fig. 3): Analysis of actual requirements, interpretation of the requirements into technical performance specifications of structures, creation of alternative structural solutions, life cycle analysis and preliminary optimisation of the alternatives, selection of the optimal solution between the alternatives and finally the detailed design of the selected structural system. The conceptual, creative design phase is very decisive in order to utilise the potential benefits of integrated life cycle design process effectively. At that phase, the design is done at system level. Modular systematics help rational design, because the structural system typically owns different parts, here called modules, with different requirements e.g. regarding durability and service life requirements.

```
┌─────────────────────────────────────────────────────────────────┐
│ ANALYSIS OF FUNCTIONAL REQUIREMENTS                               │
│ SPECIFICATION OF TECHNICAL PERFORMANCE REQUIREMENTS               │
│ CREATION AND SKETCHING OF ALTERNATIVE STRUCTURAL SYSTEMS          │
└─────────────────────────────────────────────────────────────────┘
```

```
┌─────────────────────────────────────────────────────────────────┐
│ ENVIRONMENTAL DESIGN                                              │
│          - analysis of environmental expenditures                 │
│          - checking environmental requirements and criteria       │
│          - development of the structural system                   │
│          - renewed analysis                                       │
└─────────────────────────────────────────────────────────────────┘
```

```
┌─────────────────────────────────────────────────────────────────┐
│ MULTIPLE CRITERIA DECISION-MAKING AND SELECTION OF THE OPTIMAL    │
│ ALTERNATIVE                                                       │
│      - definition of performance parameters                       │
│      - transfer of values of environmental expenditures from the analysis results │
│      - definition of weighting factors for performance and expenditure factors │
│      - multiple criteria selection between the design alternatives │
└─────────────────────────────────────────────────────────────────┘
```

```
┌─────────────────────────────────────────────────────────────────┐
│ MECHANICAL DESIGN    PHYSICAL DESIGN         DURABILITY DESIGN    │
└─────────────────────────────────────────────────────────────────┘
```

```
┌─────────────────────────────────────────────────────────────────┐
│                 FINAL INTEGRATING DETAILED DESIGN                 │
└─────────────────────────────────────────────────────────────────┘
```

Fig. 3. Process of the integrated structural design (Sarja, 1995) /4, 5, 7/.

5.3.1.2 Methods of integrated life cycle design

Design for multiple requirements and entire life cycles of buildings implies an increased number of design methods, some of which are listed in Fig. 4. Among them are some traditional methods, mainly in the area of structural mechanics and building physics. A special introduction of new methods is needed in environmental design, as well as in multiple requirements decision-making, service life design and durability design.

The principle of sustainability brings with it the introduction of life cycle methodology, and thus time as a new dimension in design calculations. Health aspects during use have to do with the control of moisture and thermal conditions, and with special subjects like unhealthy emissions from materials. In the construction phase health is related more to working conditions and safety. Design for recycling is an important tool for saving natural non-renewable resources and reducing the environmental impact.

The key issues from the environmental viewpoint are the life cycle monetary and natural economy, and the service life design. The safety and mechanical serviceability is guaranteed through the traditional mechanical design with the methods of static and dynamics. The controlled technical serviceability through the target service life is guaranteed by the durability design /8/. Health is protected by the methods of building physics including hygrothermal physical and chemical methods. The selection of final solutions between alternative structural ideas, materials and products can be done applying the methods of multiple requirements decision-making. The active reduction of the wastes in construction and renovation is possible through designing for selective dismantling in renovation and for recycling of new structural systems, components and materials.

Currently, the monetary economy and the natural economy may be in contradiction with each other, e.g. because of the pricing and the different taxes between work and natural resources. These cases then lead to valuation problems, which must be decided by the clients using their defined valuation in the framework of a society's norms. In addition to the calculated expenses, there are also factors which cannot be numerically determined. Such factors are e.g. the impact of construction on the biodiversity and the production of noise, which have to be evaluated and valuated separately by society's general rules for individual design cases.

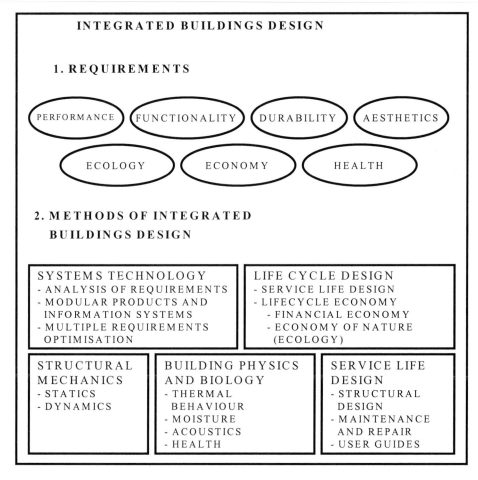

Fig. 4. Framework of integrated structural design (Sarja, 1996) /2, 6, 7/.

Introducing integrated design principles into practical design is quite an extensive process. Not only is the work of structural engineers changing, but the form of their co-operation with other partners of construction and use will have to be developed – particularly if the structural engineers' expertise is to be maximally effective at the decisive creative and conceptual phases of design. This kind of co-operation also helps clients realise the benefits of investing slightly more in the structural design. Another important change in the design will be some kind of modulation, which means separating the designing of functions, spaces and performance specifications from that of technical systems and modules. The first part of the design, which is performance oriented, will be the realm of architects and technical designers working closely together and with clients and users. The second part, which is a concretising phase, will involve closely knitted teams of technical designers and manufacturers. Because the concretising phase is often bound to the specific building concepts of contractors and their suppliers, such teams will preclude the current problem of diversified design and manufacturing processes without compromising the functional and performance requirements and other requirements for the life cycle use of the building, which are discussed above.

5.3.1.3 Factors of sustainability

The results of comparisons regarding ecological requirements, generally speaking, lead to the conclusion that differences between different materials and structural solutions of the construction phase are quite small. On the contrary, quite large differences are observable between life cycle sustainability factors of existing entire buildings or other facilities. The differences are caused by differences in the basic factors of sustainability, which are flexibility in design, buildability during the manufacturing and construction phase, changeability during use, durability by comparison with the design service life, and the recyclability of components which have quite a short service life.

In buildings the energy consumption is economically important and dictates mostly the environmental properties in the life cycle, the differences in environmental economy between different structural systems being otherwise quite small. For this reason, besides well-controlled heating, ventilation and heat recovery, the thermal insulation of the envelope is important. The bearing frame is the most massive and long lasting part of the building, and the durability and flexibility in view of functional changes, spaces and service systems are very important. The envelope must be durable and, as mentioned above, have an effective thermal insulation and a safe static and hygrothermal behaviour. The internal walls have a more moderate length of service life, but they have a requirement of coping with relatively high degrees of change, and must therefore possess good changeability and recyclability. An additional property of an environmentally effective structural system is a good and flexible compatibility with the building service system, as the latter is the most frequently changed part of the building. In the production phase it is important to ensure the effective recycling of production wastes in factories and on site. Finally, the requirement is to recycle the components and materials after demolition.

All factors mentioned above are related to the properties connected to the function and performance of the buildings. We know that the decisive factor in our society is financial economy. The budget must always stay within the agreed limits and plays a major role when decisions between design alternatives are made.

Conclusively, the most important sustainability factors in performance for structures with long target service life can therefore generally be defined as flexibility towards functional changes of the facility and high durability, while in the case of structures with moderate or short target service life changeability and recyclability are dominating. The competitiveness in sustainability between materials and structures focuses on the question of which materials and structures are able to be produced, designed and manufactured with skill and at the same cost, for the best sustainability of the building.

5.3.1.4 Modular systematics in design

In advanced building we can apply so-called modular systematics /1,2/. Modulation involves division of the whole into sub-entities, which to a significant extent are compatible and independent. Compatibility makes it possible to use interchangeable products and designs that can be joined together according to connection rules to form a functional whole of the building or in another structural system. Typical modules of a building are: bearing frame, facades, roofing system, partition walls and building service systems. The modular product systematics is firmly connected to the performance systematics of the building.

The basic description of a hierarchical building system is discussed later and is presented in Figure 5.

5.3.1.5 Recycling aspects in design

The construction of new buildings and renovation and demolition of old buildings account for about 15% of all wastes produced by the society, the largest volume being concrete and masonry wastes. The consumption of building materials can be considerably limited with effective recycling and use of by-products like blast furnace slag, fly ash and recycled concrete.

The components of the environmental profile of basic materials already include recycling efficiency, which means environmental expenses in recycling. It is important to recognise that the recycling possibilities of building components, modules and even technical systems must be reconsidered in connection with design. The higher the hierarchical level of recycling, the higher also the ecological and economical efficiency of recycling /3,7/.

5.3.2 Open and modular production

5.3.2.1 Principles and organisational aspects

Today the industrialisation of building means the application of modern systematised methods of design, production planning and control as well as mechanised and automated manufacturing processes (Sarja, 1987).

The required openness refers to the capability to assemble products from alternative suppliers into the building and to exchange information between partners of the building process and inside the consortia and business networks. The application and exchange of products, services and information nationally and internationally and the adaptation of the products and services into varying local needs and cultures is essential. For this purpose an effective international co-operation is needed in order to develop proposals for definitions, rules and models for this kind of regional and local open building, utilising global technologies and methods into local applications.

The overall system is aimed at forming an entire building from interacting items. The system thus can be defined as an organised whole consisting of its parts, in which the relations between the parts are defined by rules. The system can be a product system, an organisational system or an information system. In the open industrialised building product system, the organisational system and information system are bound together.

The central scope of open industrialisation includes the following areas:

– Demand Side, dealing with user requirements and with the introduction of the requirements into designs.
– Supply Side, dealing with the production requirements and with the linking of demand and supply.
– Building organisation and communication in building projects.

Open system building is a global framework for the building industry, including modular systematics of products, organisation and information, dimensional co-ordination, tolerance system, performance based product specifications, product data models etc., so that the suppliers serve products and service modules that will fit together.

Openness is a concept with many aspects, like:

* OPEN for competition between suppliers
* OPEN for alternative assemblies
* OPEN for future changes
* OPEN for information exchange
* OPEN for integration of modules and subsystems.

In advanced building systems we can apply the so-called modular systematics of open building. Modulation involves division of the whole into sub-entities, which to a significant extent are compatible and independent. Compatibility makes it possible to use interchangeable products and designs that can be joined together according to connection rules to form a functional whole of the building. Typical modules are: Bearing frame, facades, roofing system, partition walls and building service systems.

5.3.2.2 Modular industrialisation

Modularised systematics includes the open building product system, the open information system and the open implementation (organisation) system (Sarja, 1987) (Fig. 5).

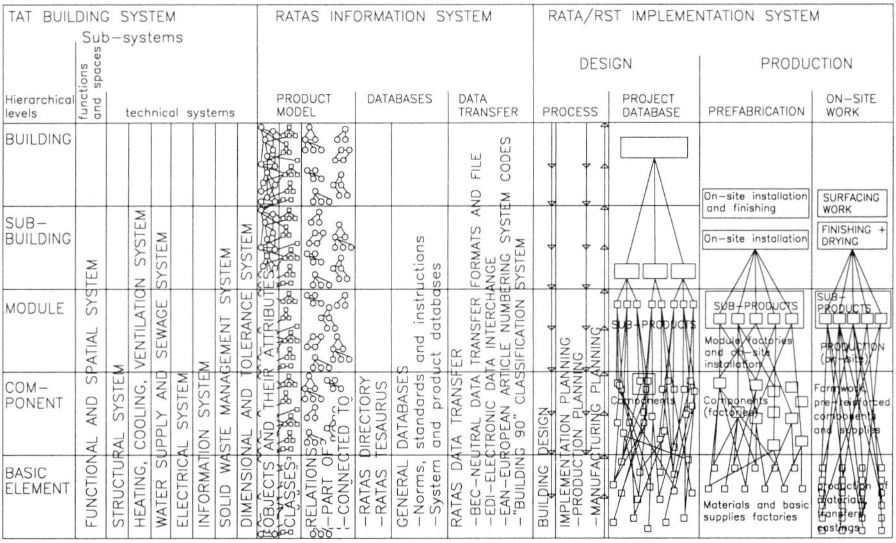

Fig. 5. The general systematics of industrialised building.

Modularisation in practice includes interaction between the business strategy of the contractor and the building process, the client oriented modularised product development, the rapid integrated order-delivery process and the management of joint surfaces between supplies and other partners. Companies and clients are building local and global (international) networks and consortia for development of building concepts and for the production itself. This will activate the continuous evolution of products and processes. As a result, advanced building technology will increasingly develop globally as an international technology development business, and will be applied locally in partnerships between technology suppliers, contractors and suppliers in close co-operation with clients and users.

At all phases of the life cycle of a building, the hierarchical modular product systematics of the building can be applied (Fig. 6). Modular systematics can be utilised in classifying different parts of the building into different classes of service life. Typically the bearing frame represents the long term parts which must be flexible to changes in the use and spaces of the building. The floors also serve as the horizontal distributors of building service systems, which is why compatibility between floors and installations is extremely important.

171

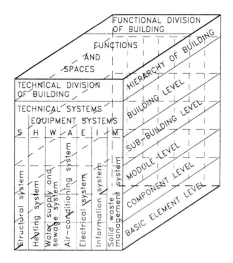

Fig. 6. Overview of the hierarchical modular building system.

The compatibility can be achieved either through flexible integration or through separation of the structures and installations. The envelope usually also has a requirement of long service life and of good maintainability and repairability. The partition walls need to be altered when the functional spaces are changed. In addition, partition walls are important locations of the building service systems. The most concentrated service distribution parts are the connection modules, which include the staircase, lift, vertical pipings and wirings, horizontal distributing connections of service systems, and possible distributed building service equipment. Building service systems HVACEIW (heating, ventilation, cooling, electrical, information and communication and waste management) are important to develop in the modular principles, especially taking into account their interaction and compatibility with the structural system. The building service functions can be distributed into the technical service systems in innovative ways, e.g. by combining the heating, cooling and ventilation with a computerised operational control system into an integral module. The systematics will be presented as model designs, alternative organisational models and applied product data models. It is important to identify and analyse productivity factors; the results can then be used to develop methods for improving productivity.

The general rules and models can be concretised into building concepts for defined consortia or networks of contractors and suppliers

(Figs. 5, 6 and 7). Each building concept can be adapted to the specific requirements of the users and society. In this way a real product development of buildings can be carried out. The building concepts can be profiled according to their properties, e.g. as low-energy buildings, ecological buildings, economical buildings, long life buildings etc.

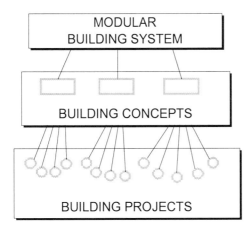

Fig. 7. Development of building concepts based on general open building systematics.

For the implementation of results into practice, companies have to develop long term strategic development projects for a 5 - 10 year period. These should include overall business strategy, product and method strategy and information system strategy. The strategy is then followed by the development of the business networks and consortia and by definition of the building concepts for them. These building concepts can be described and tested through model designs and experimental building projects. Development can be divided into three phases:

– Implementation with current products using new organisational procedures, computer applications and design solutions,
– implementation by means of partial product and method development and organisational changes, and
– implementation exploiting the entire potential for the development.

In all subareas an important task is the transfer of generic principles and models from other fields of technologies into building and to participate in applied technical research on the core areas of industrial technologies. Such research areas are e.g. the STEP systematics of product modelling and several ISO standardisation works regarding technical specification systematics.

5.3.3 Integrated information

During this and future decades information technology will continue to revolutionise working in building projects. It is important to recognise both the potential benefits of computers in all phases of the service life of buildings and the barriers in the practical use of computers. In addition, major changes must be considered even to organisations and processes in the design, manufacture, project management, use and maintenance, repair, reuse and disposal of facilities.

The modern information system is based on the rapid communication network between the partners of building projects. All communication is increasingly integrated into computers, thus integrating the design and production planning and control into the communication between partners and general databases (Fig. 8). This helps to avoid the difficulty caused by long distances. Internet is helping in international communication, thus increasing the efficiency of application of internationally most advanced technology and knowledge. Corporate and project information systems may provide a more integrated and immediate view of all aspects of a project to all partners of a construction project. Computer aided communication will have a significant impact on building processes in the future. New technologies in telepresence and virtual reality can lead to new working methods.

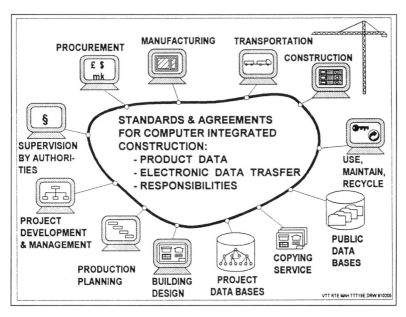

Fig. 8. Example of a computer integrated construction theme.

On the way towards real computer integrated construction (CIC), we have to solve several problems which are typical of building production. These problems arise from the complex organisational structure of the building design and construction with its very many partners and external connections with the market and society. In order to guarantee a fluent information flow between the partners of the building process, open systematisation of the data structure and data transfer is needed. In CIC, the whole information process controlling the material process is produced and exchanged between partners and stored in digital form with the aid of computers. The design database is built by designers during the design process. The database is based on an object oriented product data model of the building. In the final stage it includes all the information needed to construct and maintain the building. The system defines the hierarchical classification of the object, as well as their attributes and mutual relationships. The partners in the building process, namely the owner, the design partners, the controllers and the manufacturers, can get their basic information from the design database. Utilising their own computer aided information processing systems, all partners modify and add information, thus producing final information for the execution and control of the building process as well as for the operation and maintenance of the building. The designers employed by the manufacturer work in very close co-operation with the manufacturer, which is either a factory or a contractor. The product planning in the production unit makes possible the effective production of the building components. Through the open system, interactive data transfer between designers and manufacturers is possible. In the advanced building technology of the future, the design data will even be transferred directly in digital form into numerically controlled automated machines and robots. In the manufacture of some components like windows, doors and some structural elements, automated numerically controlled machines are already in use.

The information system consists of the following parts:

1. General databases
2. Data transfer system consisting of the following parts:
   - EDI (Electronic Data Interchange) systematics.
   - EAN Article Numbering according to European Article Numbering Systems codes.
   - Transfer Formats and File.
3. Building Product Data Model Systematics.

The general database contains building regulations and instructions, structural solutions in the form of CAD images and symbols, building product suppliers and materials, cost information, an enterprise and service index, calculation and expert programs, an electronic mail service index, calculation and expert programs, electronic mail and some other services. For electronic data exchange between partners, the system offers a definition of electronic data interchange (EDI) including article numbering codes, and definition of EDI messages. The electronic data exchange between CAD systems is defined with a neutral file approach. The principle of the neutral file approach means that the information is transformed from one system into a neutral file and again from the neutral file into the second system. The neutral file format is characterised by an extremely compact file size. The information format for the exchange is defined for the following types of data: text strings (combinations of text and pictures), information in tabular form, vector graphics, raster graphics, bar codes, knowledge and wire graphics for three-dimensional visual models.

The product data model defines the systematics for the structuring of the data for use in design, production and maintenance. In the product model, the description of a building is made using objects, their classes, attributes and relations. Information about an object's properties, i.e. its attributes, can be of many different types. In present design documents, such attribute data can appear explicitly or implicitly in drawings, in bills of quantities and in specifications. The most important attribute types are:

- Numerical value, for instance geometrical data or prices.
- Text. Strings of characters which have no meaning or internal structure but which can be transferred to different documents.
- Pictures. A bitmap, or possibly an analogue picture or even video sequence. Can be transferred to different design documents and viewed.
- Codes. Strings of characters from a predefined set of permissible values.
- Lists. Arrays of numerical values, texts, pictures or codes.

A class specifies the existence of attributes of the objects which belong to it. Each object is thus an instance of its generic class (the object has and is a relation to its class) and contains specific values for its attributes. Each object belongs to at least one class.

The open software development for design and production is a key factor for the future development of informatics in construction. Even now, most programs are still mainly drafting-oriented and do not offer effective support for construction and use of building facilities. The new generation of software will also combine the work of partners in construction and use with modern communication between them, creating a whole new era of construction.

5.3.4 Productivity through industrialisation and business networking

5.3.4.1 Productivity through prefabrication

There exists a global need for increased productivity in the building sector in order to reduce production costs. Experience from several sources shows that a rapid increase in productivity can be achieved by means of industrialisation of the building process through increased prefabrication. For example, in the last 20 years the productivity in Finland, including labour and capital productivity, has grown by 40% in the building products industry compared with only 10% in site works (Figs. 9 and 10). Labour productivity has doubled in the products industry but improved by a third on site. The productivity of capital has thus somewhat decreased especially on sites because of a current recession in building production.

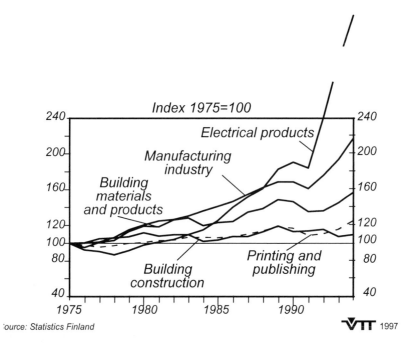

Fig. 9. Productivity increase in some industrial branches in Finland over a 20 year period /6/.

The development of productivity reflects the increase of industrialisation in building, which has caused an increase in the share of prefabrication and decreased the share of site works (Fig. 9). Industrialisation has enabled the reasonably good development of productivity in the building products industry to benefit the total productivity.

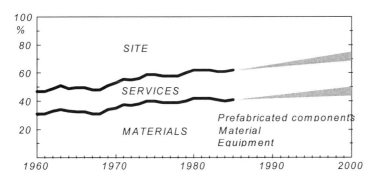

Fig. 10. Distribution of building costs in Finland /6/.

The structural change in construction has had a heavy impact on companies. In new construction, increased subcontracting has affected company structure. The total number of companies has increased and the mean size of the companies has decreased. Increased renovation and modernisation has provided employment opportunities for small firms.

Increased productivity has reflected mainly on the costs of structures, which own a rapidly decreasing share of total costs of buildings (Fig. 11). The costs of building services and finishing are increasing especially in office and commercial buildings due to the increasing number of equipment and quality of products. However, from 1990 to 1996 building costs rose by a total of only 5%, which is clearly less than the current inflation rate and less than the increase in living costs. In housing the share of building services is less – about 20% – and the share of structures still predominates.

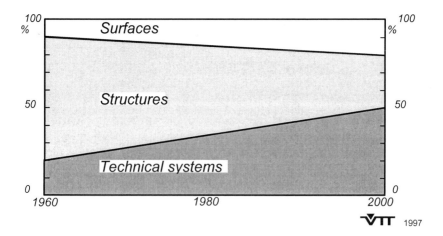

100
%

Surfaces

100
%

Structures

50

50

Technical systems

0

0

1960 1980 2000

VTT 1997

Fig. 11. Distribution of building costs in Finland between different parts of the building /6/.

5.3.4.2 Productivity through site automation

Besides industrialisation through increased prefabrication, there is another possibility which has made significant inroads in Japan /10/, namely the advanced mechanisation and automation of site processes. The goal is to create a "field factory". This concept can be either a combination of prefabrication with automated assembly on site, or an automated site manufacturing process, or a combination of both. An existing model for this concept are shipbuilding yards, along with some prototypes for other sites. In Israel, on a smaller scale, there are site robots for spraying of paint and for block assembly on sites in the prototyping phase /11/. It is probable that site automation will be increasingly applied, first on large building sites and later on smaller ones. Although initial applications will be prototypes, a wider application can be expected within 10 to 15 years. The most common type of development of site automation will probably progress in small steps, through increasing automation and computer control of site machinery and equipment. Examples of such developments already exist e.g. in cranes, shotcreting machines, drilling machines, components assembly manipulators etc.

5.3.4.3 Productivity through networking

Networking is a core principle in the modern production infrastructure (Fig. 5), where a network of contractor and specialised suppliers can be effectively utilised. The network typically includes continuous development of the products, manufacture, production and design as joint projects between partners.

The increase in use of specialised suppliers in building will increase the role of suppliers at the development and execution stages of building projects. In the network different issues and phases of the design and manufacture can be integrated into general technical development and evolution of the building production, applying the best local resources combined with international knowledge and technology. Building time schedules can be compressed into a short time period by applying an effective parallel production planning schedule, where different supplies are manufactured in parallel by several suppliers and only the assembly and finishings must be scheduled as site jobs (Fig. 12).

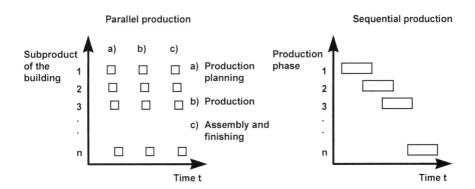

Fig. 12. Parallel production through networking between the contractor and suppliers.

Computer aided design, manufacture, project management and communication strongly undergird the networking process. Internet is an important tool in international communication, increasing the efficiency of application of the internationally most advanced technology and knowledge. Suppliers are increasingly combining services, like design and assembly works, into their deliveries. Some suppliers are specialising in technology and know-how deliveries.

The roles of partners in networking is developing toward specialisation and increased skills. Contractors are increasingly focusing skills and development into the management of building projects, while suppliers are focusing on special technologies, products and services. This transition in the infrastructure of companies has been an important factor in recent years. It has been recognised that despite the current recession in construction the number of small specialised supplier enterprises is growing rapidly /6,12/.

5.3.4.4 Productivity and quality through materials and structural engineering

It is important to recognise that managerial, organisational and design development alone cannot lead building technology to fulfil the multiple requirements facing it without the strong support of the core techniques: materials and structural engineering. The development of materials and structures will have to be done in close interaction with managerial, organisational and design development. This means that the materials and structures must be suited to mechanised and automated manufacture and tailored for different requirements. This interaction between different techniques and skills is described in Figure 13. The development of materials and structural techniques will happen mainly at the hands of specialised suppliers in interaction with the development of manufacturing technique, and by applying integrated life cycle design for fulfilment of the multiple requirements of each system, module or component. Contractors will concentrate on the link between the client, the building as a final product, and the project planning and management phases, while applying appropriate multiple requirement analysis, optimisation and decision making.

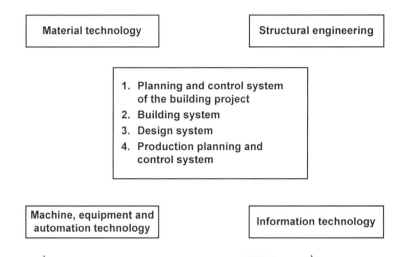

Fig. 13. Basic technologies and their interaction in building

Concerning materials and structures, new basic knowledge will be needed especially regarding hygrothermal behaviour, durability and service life of materials and structures in varying environments. Such knowledge will have to be put into practice through standards and practical guides. The creation of new types of materials and structures, in which the properties can be tailored separately for each specific need, is of vital importance. Both strong and soft materials and structures are needed, depending on the specific life cycle requirements in each application. Another challenge for materials engineering is the effective recycling of building and demolition wastes. The construction area is producing about 10% of all wastes in society. There are already good examples of how these wastes can be re-used in construction. As a major consumer of materials, construction can apply many by-products and recycling materials from industry and general consumption, but new creative innovations and applications are still needed for meeting those targets.

REFERENCES

1. Sarja, Asko, Principles and solutions of the new system building technology (TAT).Technical Research Centre of Finland, Research Reports 662. Espoo, Finland, 1989. 61 p.

2. Sarja, A. & Hannus, M. Modular systematics for the industrialized building. VTT publications 238. Technical research centre of Finland, Espoo 1995. 216 p.

3. Sarja, A. Towards the advanced industrialized construction technique of the future. Betonwerk + Fertigteil-Technik 4/1987 pp. 236-239.

4. Sarja, Asko, Methods and Methodology for the Environmental Design of Structures. RILEM Workshop on Environmental Aspects of Building Materials and Structures, Technical Research Centre of Finland, Espoo, Finland, 1995. 5 p.

5. Sarja, A., Prefabrication in relation to sustainable building. Symposium Report: Prefabrication Facing the New Century. Concrete Association of Finland, Helsinki, October 1997. Pp. 143 - 146.

6. Well-being through construction in Finland 1997. Technical Research Centre of Finland (VTT), Building Technology. Vammalan kirjapaino Oy, Vammala, 1997. 32 p.

7. Sarja, A., Framework and methods of life cycle designs of buildings. Int. Congress "Recovery, Recycling, Re-integration", R`97, Geneva, Switzerland, 4-7- February, Vol. VI, pp. 100 - 104.

8. Sarja, Asko & Vesikari, Erkki (Editors), Durability design of concrete structures. RILEM Report Series 14, Rilem TC 130 CSL. E&FN Spon, Chapman & Hall, 1996. pp.

9. Sarja, A. Guide for integrated life cycle design. First Draft 21. 05. 1997. Manuscript for the report of RILEM TC EDM / CIB TG 22.

10. Study on innovative and intelligent field factory. International Robotics and Factory Automation Center. IMS Promotion Center, Tokyo, March 1997. 33 p.

11. Warzawski, A. Industrialization and Robotics in Building. A Managerial Approach. National Building Research Institute Technion-Israel Institute of Technology. 1990 New York. 466 p.

12. Konjunktur`98, Gute Grundstimmung. Wirtschaftswoche Nr 52, 18. 12. 1997, Pp. 30- 31.

6

Open building development in Japan

Seiji Sawada

D.Eng., Architect (Japan)

6.1 DEVELOPMENT OF HOUSING INDUSTRY IN JAPAN : 1960-1990:

During the three decades, 1960-1990, Japan built up a modern industrial society. During the first decade Japan achieved high productivity. Japan realized a high standard of modern living, and in the second and third decades the Japanese people enjoyed what was accomplished in the first decade.

(1) Regarding the productivity of the building industry, the housing floor space produced per worker increased from 100 sqm to 300 sqm, according to a report from the Building Center of Japan, BCJ. This figure has not changed since then, while other industries increased their productivity.
(2) The BCJ report indicates also the introduction of new building technology to the traditional timber construction, namely, house construction using also industrially produced building components and prefab house production. During the years 1960 - 1990, housing construction using industrial building components and prefab house production tripled their productivity.
(3) From the year 1970 until the year 1990, the standard of living - in terms of per capita GDP - grew sixteen times in Japan. This is the progress of an industrial society.
The productivity in other leading industries, such as machinery and electronics, continued to make rapid progress, whereas that in the building construction increased no more.

Considering that a large part of housing production consists of labor cost, the following points can be made about the relationship between the product price and the buying power of the user :

- The price of housing industry product increased almost at the same rate as that of the user's income.
- The price of housing industry products became increasingly more expensive that other products, such as cars and home electric consumer goods.

(4) Let us look at the transformation of the housing production industry. I will show three charts, which illustrate the content of housing production comprehensively in the years 1963, 1973 and 1985 (Fig. 1.). Each square represents the number of dwelling units produced in the years 1963, 1973 and 1985. The total volume of housing production is divided horizontally to show the use of traditional methods and the use of new production methods. The left side is the traditionally produced housing and the right side is the housing produced by new technologies and methods. The total volume is also divided vertically to show the different housing types. The upper side shows that of detached one family of tenement type houses and the lower shows high-rise collective housing types.

- The volume of total housing production increased from 690,000 units in 1963 to 1,910,000 units in 1973 and then deduced to 1,240,000 units in 1985.
  It is clearly seen that the production of housing in the detached one family or tenement houses, where the home builder used traditional technology had also decreased gradually.
- At the same time these decades were a period of population flow to the large cities in Japan. It was also a period of new technologies such as reinforced concrete and steel and reinforced concrete methods being introduced to build high-rise and other collective housing. The introduction of these new technologies became available only for general contractors. They became dominant in the housing industry during the period. (Re. 1)

1963

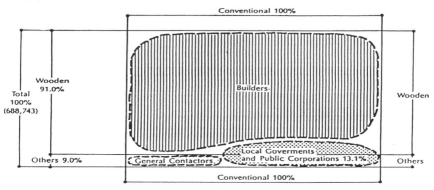

1973

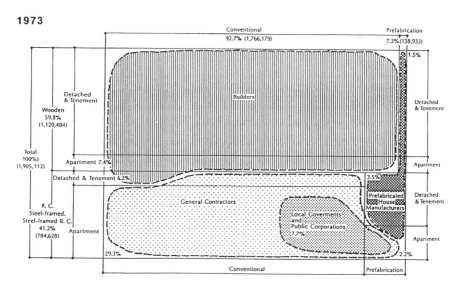

1985

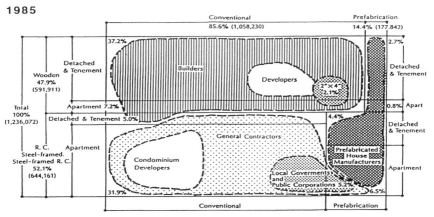

Fig. 1. Comprehensive content of housing production in the years 1963 - 1985.

187

- Along with the appearance of the new leader in the area of "collective housing" production, a new housing industry appeared in "one family or tenement housing" production, that is the prefab house industry. At the same time that the one family or tenement housing production increased - from 7.3% in 1973 to 14.4% in 1985. - a number of prefab house companies were established in Japan. Large investments were made by the already established companies to build highly automated production plants. While general contractors made efforts to industrialize the housing production on construction site, the prefab house industries made efforts to develop production systems in factories.
- In the light of the development of the building industry, in Japan from 1960 through 1990, the following points can be called characteristics specific to Japan.
  (a) In the course of establishing a modern industrial society in Japan, the population rapidly concentrated in big cities.
  (b) The collective housing form - high-rise housing and others - has increased in response to this urbanization, from 10% to 50% of housing production.
  (c) One family or tenement type houses managed by home builders according to traditional
  (d) technology have also increased.
  (e) Three new housing organizations have appeared in the housing industry :
    - General contractors for the collective housing.
    - Prefab house companies for one family or tenement houses.
    - Building component manufacturers for windows, doors and home appliances.

6.2 RECENT DEVELOPMENT OF HOUSING INDUSTRY IN JAPAN : AFTER 1970:

(1) The above described new housing industry was established by the beginning of the 1970s. During the 70's housing statistics indicated that the "one house per one household" level was achieved. People started to enjoy a high standard of living :
  * During the 60's the following household items became status symbols :
    - Television, refrigerators and washing machine.

188

* During the 70's three new items called the "3C"s became status symbols :
  - Color television, car and cooler, i.e., air conditioner.

(2) The rapid development of modern industries in the capital-intensive market in Japan also occurred in the housing industry. New technologies and industry organizations have been developed to meet the diversified consumer needs.
  - During the 70's, there were 10 different product types per prefab house manufacturer. Afterward, this grew 10 times, to one hundred types. (a study by Prof. Matsumura, University of Tokyo)
  - Building component suppliers made a large number of items. The product catalog of TOTO sanitary appliance company was only ten pages in the year 1970. This catalog grew to one thousand pages in the year 1990. (a study by Dr. Iwashita, Atias Corp., Tokyo)

(3) Capital concentration was needed in order to transform business corporations during this period.
  - Vertical grouping of subcontractors was done significantly at five to ten major general contractors.
  - A number of small and weaker prefab house manufacturers was absorbed by about ten major prefab house manufacturers.

(4) Most prefab house manufacturers adopted planning, manufacturing and marketing technologies from car and home electronic product manufacturers. The prefab industries were losing characteristics of the construction industry and taking up characteristics of the manufacturing industry.
  The general contractors expanded their business lines to overseas market and urban development project markets. In the housing market some of the major general contractors started a business of taking care of the people living in housing which they designed and built.

(5) Two important elements have been supporting this transformation of the housing supply corporations :
  - A new business organization concept was developed to manage the whole process from market research, planning, design, production and maintenance of the products and services.
  - Also they developed computer-based automation of information handling in this work process.

6.3 EFFORTS MADE TO DEVELOP OPEN BUILDING IN JAPAN :

(1) During the 60's, Support- and Infill-technologies and production organizations were developed.
- The Japan Housing Corporation, a semi-public housing developer established in 1960 to supply housing for large city inhabitants, took the initiative of developing Open Building components which were to be fitted in their concrete Supports.
- Prefab house companies, who started their technological systematization on traditional timber house technology, developed inner wall systems independently and kitchen and sanitary systems in collaboration with building component manufacturers.

(2) In the 70's, efforts were made to develop the concept of distinction and integration of Support and Infill.
- The building system development projects were initiated by the Japan Housing Corporation and the public Housing Corporations on the local authority level. For example SPH and NPS aimed to develop systems based on standard floor plans.
- The Japan Housing Corporation - Nihon Jutaku Kodan - conducted an interesting experiment project called KEP - Kodan Experimental Project - in order to develop interfacing rules between Support and Infill for housing design and production with high flexibility of house form and floor plan.

(3) Also in the 70's, a new trend was born in the housing industry with the achievement of a higher standard of living. It is called "dweller-initiated home building" which aimed to withdraw from "highly industrialized" and reorganize "hand made" housing patterns : this is called "cooperative housing." This is a new direction in Japanese housing, even though it is still small. It is interesting that this "user-participation" housing pattern was influenced by the Open Building theory developed by SAR - Stichting Architecten Research - in the Netherlands.
- Citizen-initiated housing movement is observed in OHP projects in Tokyo and Tojuso in Osaka.
- Housing projects with user participation are also being developed by the Japan Housing Corporation, such as "free plan housing" and "menu supply housing."

(4) After above experimental projects, studies of two important housing systems were made during the 80's which are bases for future Open Building development in Japan.

- Century Housing System is a classification of building parts in accordance with their life span.
- Two-step Housing Supply System : Study of building deregulation to enable the separation of supplying Support and that of Infill.

(5) The efforts made in Japan to develop the Support and Infill concept, and technologies and techniques to manage Open Building design and production will be summarized as follows :

- Technology and industrial capabilities are accumulated in the general contractors and the prefab house companies.
- The expression of the three level design method, as proposed by the SAR, which is to determine Infill, Support and Urban Tissue based on agreement in families, neighborhoods and communities, has not yet been fully realized, except for in a few urban renewal projects.
- Support and Infill technologies accumulated in large general contractors and prefab house companies are used only in housing production activities for the market segment which each corporation manages. This technology system should be called not "Open" but "closed" since it is open only within each corporation.
- This type of Support and Infill housing was developed in Japan due to the following reasons :
  - (a) The volume of production for each corporation is large enough to enjoy the benefits of systematization.
  - (b) Computerized manufacturing can supply products of great variety with high efficiency.

6.4 FUTURE DEVELOPMENT OF OPEN BUILDING IN JAPAN :

(1) The "High-rise Housing System Development Project" initiated by the Ministry of Construction is now in progress. From the present state of this government-industry-collaborate project we can learn some matters which are important to predict how the Open Building in Japan will be developed in near future and who will be the leader of the development.

- Main background of the project : shortage of skilled workers in large cities, not well developed design and production systems in this housing area, and therefore strong uniformity of design.
- Targets to be reached in the project are : systematization of planning, design, costing, construction and maintenance works,

technology development to improve productivity and quality of product, new design method which also covers community facilities, and Infill systems which can fit the new consumer-ruled market.

– Regarding the above-mentioned "systematization," the Ministry of Construction aimed to put together potential Support technologies from the general contractors and Infill technologies from the prefab house manufacturers. It is interesting that middle-size general contractors are involved more deeply in this long-term innovation project than major general contractors.

(2) There are significant and rapid changes in the housing market which will be a basis for a long time.

– Society and the economy in Japan have reached a mature level. No high rate of economic growth is expected during coming decades.

– In response to this change, "competition and harmonization" is being discussed in business corporations in general and in the housing industry as well.

– The diversification of the housing market and price competition require the "down-sizing" of corporate organization and the "decentralization" of decision-making which are also needed to keep beneficial business.

(3) Who will be the leader of future Open Building development in Japan?

– The most important requirements for the leader of the development should be the capability to manage housing systems adaptable to market dominated by the consumer and the capability to manage the planning, design and construction of a long lasting urban environment. It can be said that the former exists in the prefab house companies and the latter in the general contractors.

– It seems however that these potential industries have not yet noticed this principle. Both are still spending time and money to realize new system for their market segments. They do not recognize that their capabilities are supplemental, that integration would be more effective.

– The general contractors do not need large investments because they do not have large production plants. The house makers however, need large investment for production plants and sales offices, which should be paid back within a short period. Under the condition of present low-rate of growth of the economy, this return on investment will take a longer time.

- If the general contractors will take the initiative of development and coordination of the rules for interfacing between Support and Infill, and the time they need is not very long, they can contribute to the further development of Open Building in Japan.

(4) What issues are to be discussed for the further development of Open Building in Japan ?

- Industrialization became international issue in the last three decades, as it is quite clear in the car and electronics industries. Building industrialization should also develop in the same direction. In order to conduct this, it is necessary to organize opportunities for the related experts to discuss the building industrialization from an international view point.

- Housing is at the same time a very local issue. Not international but to be based on local context. This means that the housing activities - from planning, design, construction and maintenance - in industrialized societies should be carefully programmed so that the projects can best enjoy the "internationalized" industrialization, and also contribute to maintaining the local culture and economy.

- For this purpose the area of Open Building theory used to understand housing as process will become more important than before. Greater efforts should be made from now on in Japan, such as those started with the experimental project "NEXT 21."

- The most important aspect in the "Open Building concept," which is presently demanded, is that for "sustainable development." The distinction of Support and Infill in the concept will be a good basis for the "sustainable housing building," because Infill is the building part enabling a short-term return on investment, and the Support part of building recovers investment over a long period.

- In this regard, a comprehensive view should be introduced into housing research in Japan, since the large housing estates, which were developed during 60's, are no longer adaptable to the life styles of present inhabitants. Hot discussions have started if they are to be demolished and replaced new buildings? Or should the life of the inhabitants be maintained and revitalized through the renewal or renovation of that built environments? We have to find out new solutions for this very complex problem of "sustainable development." It is therefore necessary to discuss if the adoption of Open Building concept will be effective or not, as the whole society of Japan has no longer the capability to make such large investments like those of last three decades.

6.5 CONCLUSION :

In the light of long-term Open Building development in Japan, it can be said that the Support/Infill distinction in terms of restructuring the building industry to Support- and Infill-industries will be necessary. In order for this new building industry to respond to the social changes expected during the coming decades, more thoughts about the "housing process" - project management - will be urgently required. This new housing industry concept which consists of Support-, Infill- and "Open Building-project management" industries will lead a healthy development of the Japanese housing industry. As this further development will require quite comprehensive thinking, it is necessary to have well organized opportunities to discuss these issues with experts in foreign countries.

Re. 1 : General Contractors :

"General contractors" in Japan are to recognize as companies where design, construction and project development are integrated. This integration is probably attributable to a sense of "confidence" between the client and the contractor. This is particularly true for private projects, while in public projects designs are controlled by the engineers of clients. There are ten majors which employ about 10,000 architects and engineers.

Tab. 1 : Outline of the The Open Building Development Projects :
SPH (System for Public Housing)
Initiated by the Ministry of Construction for the housings subsidized by the Ministry of Construction and of Finance, and developed by the Japan Housing Corporation, the Prefectural Housing Corporations and the Local Authorities.
NPS (New Planning System)
About five years after SPH project in the 1970s NPS project was organized in almost same concept but for the housings which became larger in response to growth of living standard.
KEP (Kodan Experimental Project)
Initiated by the Japan Housing Corporation in order to develop interfacing rules between KJ Infill components and Supports, participated by the KJ component manufacturers.
CHS (Century Housing System)
Initiated by the Ministry of Construction, participated by general contractors and component manufacturers.
High-rise Housing System Development

The Ministry of Construction organized a technology development proposal competition. About twenty groups of housing industries - general contractors, subcontractors and component manufacturers were selected and are now working in a technology development cooperative.

NEXT 21 Project

NEXT 21 is an experimental future-oriented housing project with 18 dwelling units. The project is intended to anticipate the more comfortable life that urban households will enjoy in the 21st Century. The project explored a number of methods for constructing urban multifamily housing. It was conceived by Osaka Gas Company and the NEXT 21 Construction Committee developed the basic plan and design. It has objectives; to use resources more effectively by systematized construction methods, to create a variety of residential units that meet the demands of a variety of households, to introduce natural greenery to urban environment, to treat every waste and drainage within the building and to ease the burden on the environment, and to use energy efficiently by such means as fuel cells. The project will last for decades.

REFERENCE :

A comprehensive report on present situation of Japanese construction industry: Changing Construction Markets of Japanese Companies, S. Sawada, Raketamisen Linjaukset, 1997, Tekes / Finland.

7

Industrialised building: a review of approaches and a vision for the future

Frank de Troyer
Professor (Belgium)

7.1 INTRODUCTION

"Prefabrication and modular co-ordination", "Dimensional co-ordination", "System building", "Industrialised building", ... Is there a kind of fashion in the terminology describing approaches aiming at an efficient production in the construction sector? Probably it is more than just a fashion, it reflects changing viewpoints over the years. Those shifts in approaches can be illustrated by the change of the name for the Working Group 24 of CIB.

The double name "International Modular Group/CIB W24" was commonly used up to the beginning of the 80's (ref. 1). This reflects the origin of the group as described in 1974 (ref. 2, p. 3)

> *"After the completion of EPA Project 174, in 1961, a number of individual participants in the project decided to continue international co-operation on modular co-ordination, on a non-governmental basis. They accordingly formed the International Modular Group (IMG) and co-opted members from all over the world. In the meantime, work on dimensional co-ordination in building had started also in wider international technical circles, within the framework of the International Council for Building Research, Studies and Documentation (CIB). In order to be recognized as an international working commission of the CIB and to avoid duplication of work, the IMG slightly amended its statutes and became CIB Working Commission W24."*

The EPA Project 174 was an intergovernmental project initiated by the European Productivity Agency (EPA) of the Organisation for

European Economics Co-operation (OEEC), initiated in 1954 and leading to a first report in '56 and a second in '61 (ref. 3 and 4). In the beginning the focus was very much on "modulation" and "prefabrication".

In the seventies within CIB and ISO a lot of discussion was going on between two groups:

- one group stressing the importance of cosidering "building nodes", "boundary conditions", "modular or sub-modular increments or decrements"
- the other group advocating to keep things as simple as possible, as for example in the "Condensed Principles of Modular Co-ordination" published in 1964 (ref. 5) and reflected in "ISO 2848 Modular Co-ordination - Principles and Rules".

The first group proposed to use the term "dimensional co-ordination" instead of "modular co-ordination". This discussion is reflected in the name of the CIB W24 commission as mentioned in the compendium of 1983 (ref. 6, p. 9): "W24-IMG (International Modular Group)-Dimensional and Modular Co-ordination".

At the end of the 80's Gy. Sebestyén (Secretary General of CIB) was thinking of renaming the working group "Open System Building".

In 1995, stimulated by the present Secretary General Wim Bakens, the group was reactivated and adopted its present name "Open Industrialisation in Building".

In the following points the evolution in concepts underlying those shifts in terminology will be elaborated.

7.2 "OPEN BUILDING SYSTEMS" AND "CLOSED BUILDING SYSTEMS": A SEARCH FOR A CONCEPTUAL FRAMEWORK

7.2.1 Background

The concepts "Closed Building Systems" and "Open Building Systems" are widely used in the context of industrialisation of the construction sector. There was a lot discussion about concepts and terminology over the years (ref. 7 - 15).

The easiest way to explain the notion "Open System Building" is to start with the term "Closed Building Systems". The term "Closed Building System" is used to describe the approach of the building sector after the war 1940-1945 in Europe. The term "closed" was added in the sixties and has a negative connotation.

The term "Open System Building" is a neologism for an alternative approach for those "Closed Building Systems". The easiest way to describe this alternative is to consider two intermediate steps between the two extremes "Closed Building Systems" and "Open System Building" (Fig. 1) :

- "Semi-Open Building Systems"
- "Open Building Systems".

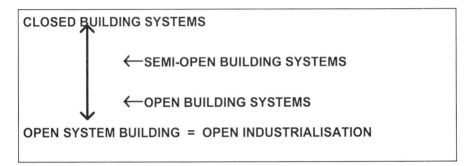

Figure 1

7.2.2 Closed building systems

Following the approach classified in the sixties as "closed building systems", first a project is designed and secondly subdivided into a number of building parts which will be produced especially for that project (Fig. 2A). In all cases a lot of components are bought on the market. Therefore one should take care that certain dimensions of the self-produced building parts fit with these other components

example: An apartment-block is divided during the design process into concrete wall- and floor plates. These concrete elements are designed in such a way that the connections between them are possible. At the same time one should take care that the openings in the wall elements are co-ordinated with the windows or doors one wants to use. Also the sanitary equipment should fit, etc...

Disadvantages of this approach are:

- only a limited choice between a few types of houses for each project (otherwise production series are to small)

- monotony of large sites
- monopolistic positions for producers and large scale contractors.

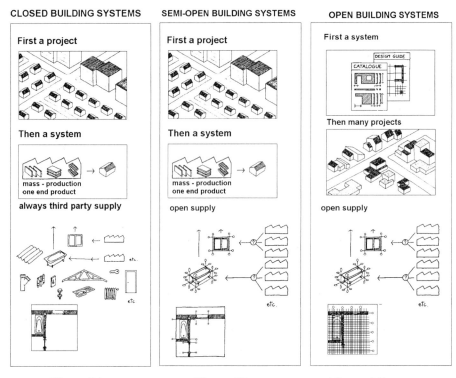

| **Figure 2A** | **Figure 2B** | **Figure 2C** |

7.2.3 Semi-open building systems

In the case of "Semi-Open Building System" one also starts from a specific project, but by respecting a certain dimensional discipline the elements produced for that project will fit with components of many independent manufacturers (Fig. 2B).

example: *It will be possible that windows and sanitary equipment of different producers can be incorporated in the above mentioned concrete buildings.*

7.2.4 Open building systems

The term "Open Building System" is used when the starting point is totally different. One does not start from a specific project but one develops a set of elements which can be combined in many ways (fig. 2C). So this term can be used when the two following conditions are fulfilled:

- with a (limited) number of elements which are part of the system, an unlimited number of different projects can be realised. There will however always be a set of constraints for the design of those projects. The "Design guide" will inform the designer of the rules he has to follow (grids, positioning-rules, requirements for stability,...). The "Element Catalogue" provides an overview of all the possible elements of a system.
- the building elements of that "Open Building System" are co-ordinated with components of different (beforehand unknown) producers. This is the condition already necessary for a "Semi-Open Building System".

In many situations not all the components needed for a project are elaborated by the same firm, only a subset. This might be called "Open building sub-system".

e.g. sub-system for windows, for kitchen equipment, for partition walls,...

The distinction between "Open building system" and "Open building sub-system" seems to be rather theoretical, because in practice it is impossible to define all the components that should be provided in order to consider a system as "complete".

7.2.5 Open system building or open industrialisation

The term "Open System Building" is used to describe a global approach for the construction industry (Fig. 3). Maybe it is better to use the term "Open Industrialisation" for this concept since the difference in terminology between "Open Building System" and "Open System Building " is only limited but the approach is completely different.

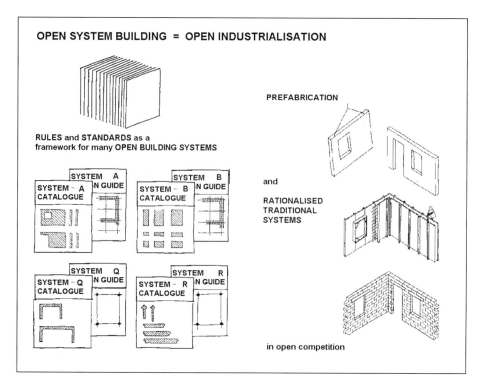

OPEN SYSTEM BUILDING = OPEN INDUSTRIALISATION

RULES and STANDARDS as a
framework for many OPEN BUILDING SYSTEMS

PREFABRICATION

and

RATIONALISED
TRADITIONAL
SYSTEMS

in open competition

Figure 3

Following conditions must be fulfilled:

- it must be possible to develop different "Open Building Systems". Due to the general rules governing those systems (concerning dimensional co-ordination, compatibility of performances,...), the decision to choose one specific "Open Building System" can be postponed during the design process as long as necessary in a specific situation;

  example: During the design process on has decided not to use a bearing wall structure but a skeleton. Of course at that moment systems making use of bearing walls are excluded, but the competition is open between system based on concrete or steel, between "column and beam" -systems and "portal frame" -systems, ea. ...

- the same building parts and the same "Open Building Sub-System" can be integrated in as many different "Open Building Systems" as possible

- the building parts of the different "Open Building Systems" can be exchanged as much as possible between such systems.

The concept "Open Industrialisation" is not only incorporating prefabrication, but also rationalised traditional methods like systems for in situ shuttering, systems for rationalised bricklaying (including prefabricated lintels, ea. ...)

The term "open" has different meanings for different authors. Even if they have the same aspects in mind the importance they give to each aspects differs. In short one may conclude that systems should be:

- "Open" for different assemblies in order to provide different lay-outs in line with a multitude of individual needs and preferences.
- "Open" for future change. If certain parts of a building fulfil basic requirements (stability, fire resistance, acoustic insulation, ...) they will have a long technical lifetime. They should be conceived to be adaptable for future use. Different policies are possible: alternative use of existing building, changing of parts that can be replaced with minor costs, expandability, ...
- "Open" for the integration of different sub-systems and third party components
- "Open" for the information exchange between different actors
- "Open" for competition between different suppliers.

7.2.6 Conclusions

Probably it should be stressed that completely "Open" systems in (line with all the aspirations expressed above) will never exists, like completely "closed" systems (without thinking of reusing components in a next project) have never existed either.

It is just by reducing the variety of components of a system (and thus the freedom of designers) that a more efficient production is possible. Those limitations should be described explicitly. An important aspect of this reduction of variation is the dimensional variation with all the consequences for grids, rules for positioning of components, rules for dimensioning of components,...

During the last decades an important evolution took place in the field of dimensional co-ordination. This is briefly described in the next point.

7.3 DIMENSIONAL CO-ORDINATION AND THE "BUILDING NODE"

7.3.1 The "building node": an old problem

7.3.1.1 Modular co-ordination only based on 1M ?

In figure 5A a grid of 1M by 1M is used to define the position and dimensions of the "co-ordinating spaces" of all the components shown in that projection.

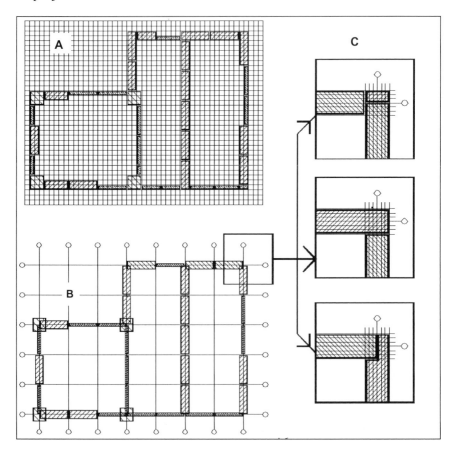

Figure 5

A "co-ordinating space" is, following ISO 1791-1973 (ref. 16), "a space bounded by co-ordinating planes allocated to a component including allowances for tolerances and joint clearance".

The surfaces of "co-ordinating planes" are perfectly plane and intersect (in our examples) under square angles. By using those theoretical volumes all dimensional deviations are excluded.

In this text, for a better visualisation, not the "co-ordinating spaces" are drawn, but components with a certain margin. The size of this margin is not important in this analysis.

7.3.1.2 Modular co-ordination based on multi-modules and/or non-modular thickness

In figure 5B a multi-modular grid is used to define the position of the "co-ordinating spaces" of components. In this sketch all the components have the same length and thickness as the corresponding components of figure 5A. All of them are positioned in such a way that their centre-line coincide with the grid line of the multi-modular grid. At several points, the "co-ordinating spaces" are overlapping. In those cases there are different solutions (fig. 5C).

The choice for one solution or for an other is made, when one proceeds in the design process: when making the first sketches the thickness of components is not jet considered, but later on decision have to be made (fig. 6).

- interruption of the multi-modular grid over a certain zone
- boundary planning (face-line) instead of axial planning (centre-line)

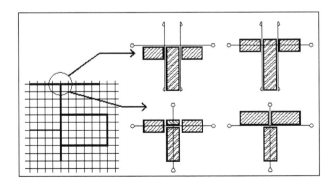

Figure 6

Only in very exceptional cases, due to a lucky coincidence of a certain lay-out, all the problems of overlapping "co-ordinating spaces" will be solved with those two measures. For many components the length of the "co-ordinating space" has to be increased or decreased. This is illustrated in a rather abstract way in figure 7.

On top of this figure there is a schematic representation of two horizontal sections of a part of a building and at the bottom of two vertical sections. The two horizontal sections are derived from the same sketch, as are both vertical sections.

Both drawings at the right hand side are elaborated on the basis of a continuous grid and at the left side on the basis of an interrupted grid.

All the distances between grid lines are multiples of 1M except in the case of those interruptions. In those cases the size of that zone between the grid lines is equal to the thickness of the "co-ordinating spaces" of the components that will be placed in that interruption.

From this figure it is even more evident how the "deviations because of the building node" can be shifted to another group of building components by choosing another grid. One has to ensure by the choice of the grid and positioning rules that "deviations because of the building node" will occur as exceptionally as possible. If they cannot be avoided, they should be shifted to those components where there can be catered for with the least cost.

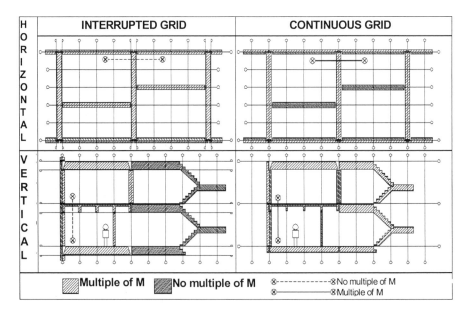

Figure 7

The problem of "the building node" occurs as well in vertical as in horizontal section.

If for the thickness of the "co-ordinating space" only multiple of 1M or sub-modules are accepted, the increments or decrements due to "the building-node" are also restricted by 1M or sub-modules.

7.3.1.3 The problem of "joints" and "building nodes"

As a conclusion one can say that there is a clear and important distinction between the problem of "joints" and the problem of "building nodes".

If length and thickness of "co-ordinating spaces" are restricted to multiples of 1M, only joints have to be considered. The size of joints is in the range of some millimetres to some centimetres.

In the case multi-modules are used in combination with modular or sub-modular thickness for "co-ordinating spaces", on top of joints also "building nodes" have to be considered. The last are in the range of some centimetres to some decimetres.

7.3.2 Modular co-ordination in the fifties and the sixties

It is impossible to prove in this short text that the "problem of the building node" was the focal point of the discussion between IMG/CIB W24, ISO TC59/SC1 and ISO TC 59/SC5.

One group estimated that this problem was only a minor one. International standards should not try to provide solutions for "building nodes" but should only mention the problem. International standards should be limited to general rules. If some countries are convinced that this problem should be tackled in detail, it should be done in national (or regional) standards.

The other group was convinced that not giving proper attention to the "problem of the building node" prevented the breakthrough of modular co-ordination.

7.3.3 Dimentional co-ordination in the seventies and the eighties

On top of the already existing dispute a new approach was developed in France, The Netherlands and countries of the DACH-group.(**D**eutchland, **A**ustria, **C**onfederatio **H**elvetica) Components where positioned in the centre of a multi-modular grid (e.g.. 3M x 3M). At "building nodes" there are two possibilities:

- adding a small component
- providing a longer component

In The Netherlands this is called "sluitstuk" or "sluitelement"(NEN 2880) In France it is called "espace d'adaptation" and different solutions for this space are elaborated.

As shown in figure 8 the length of the "co-ordinating space" of a component can be expressed in the following way:

- n times the multi-module plus twice the "espace d'adaptation".
- (n + 1) times the multi-module minus the thickness of the "co-ordinating space" of the other component.

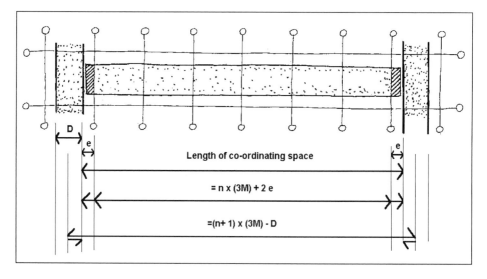

Figure 8

The advantage of this new approach is that the problem of the "Building Node" is tackled in detail, the main disadvantage is that the problem of the "Building Node" is mixed up with problems of joints.

The confusion was increased because the two groups were using the same term ("co-ordinating space) for two different concepts : a volume including at all sides a margin for joints or a multi-modular envelope that will be penetrated in certain cases of nodes (fig. 9).

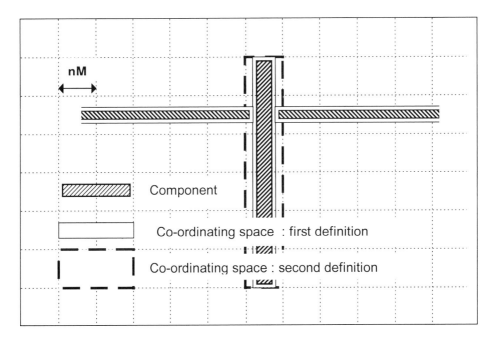

nM

Component

Co-ordinating space : first definition

Co-ordinating space : second definition

Figure 9

7.3.4 Conclusion

At the meeting of ISO TC59/SC1 in May 1996 in Leuven, following the CIB W24 meeting, an agreement was reached on two points:

- the selection of multi- modules in function of a specific application
- the acceptance of the principle of the "Building Node".

Still at this moment a group, disappointed of the lengthy discussions on dimensional co-ordination, believes that the "Building Node" will be solved on the spot and is not in favour of a detailed description.

The other part hopes that the acceptance of this principle will open the possibility for automatic or semi-automatic positioning of components in CAD-programmes. The description in great detail of the "Building Node" can even create the possibility that computers search automatically for the optimal layout.

7.4 "OPEN BUILDING SYSTEMS" WITHIN THE CONTEXT OF "OPEN INDUSTRIALISATION": SOME BELGIAN EXPERIENCES

7.4.1 Introduction

The evolution of "System Building" can be illustrated with examples from different countries. Each context has specific characteristics: climate, available raw materials, organisation of the construction sector, preferred dwelling typologies, average size of projects... In that sense every illustration will be coloured with a local palette. In this text some general trends will be illustrated with examples of the author's context.

7.4.2 Systems for housing based on large concrete panels

In the 50's and 60's typical projects making use of prefabricated concrete panels were (for the Belgian context) large scale social housing estates. They consisted of medium-rise apartment blocks as well as of one- or two-story individual units. The monotony of the end result is typical for this approach since one tries to reach large production series of the same element.

7.4.3 A flexible production tool for large concrete elements

As opposed to this "closed" approach, an "open" approach was developed (ref. :18-19). All the elements needed for the system (external walls, internal walls, floors) could be produced in a convertible moulding system. The open exchange of products between different suppliers became possible. The same rules allow the construction of those designs as well in rationalised traditional block work as in "in-situ-shuttering" systems. An open competition was created.

A prerequisite for the elaboration of this "convertible moulding system" was:

- only one height for the walls
- only one thickness for the wall elements
- the length of walls may be changed in steps of 60 cm. Therefore the sketch design should be made on a 60 cm by 60 cm horizontal grid

- Window and door openings should be positioned in horizontal projection on a 30 cm by 30 cm grid partly coinciding with the previous grid
- In vertical section floor elements are situated in a 20 cm zone (boundary position vis-à-vis lowest reference line).

Even if one accepts those constraints, several options are open as far as the detailed position of the wall vis-à-vis the grid is concerned. Analysing the feasibility of the different proposals was not easy. There was a lot of discussion as far as the technical advantages and disadvantages were concerned. But there was, at first, much more disagreement concerning the cost consequences.

In a following phase an in depth analysis of the cost of the different elements was elaborated. Of course there were differences between the different factories, but the same general trends were observed. They are summarised in figure 10.

The cost per m^2 of walls (excluding the openings):

- increases hyperbolically if walls become shorter
- increases hyperbolically if the percentage of the opening increases
- is almost independent of the number of times the same element (same size & position of openings) is used (see ref.: 19-20).

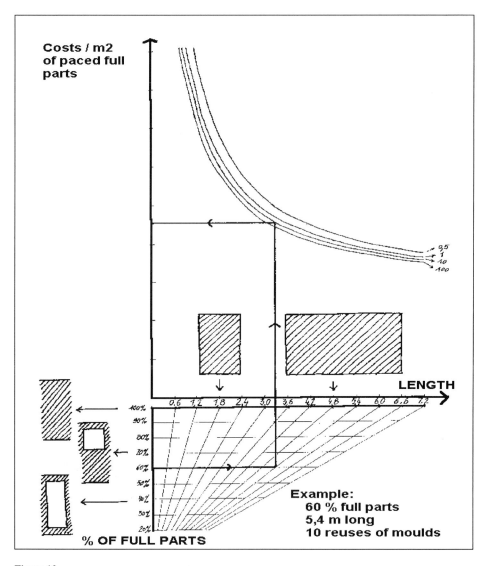

Figure 10

As a consequence of this it becomes clear that short elements with large opening are (per m² of wall) extremely expensive. Most manufacturers of those elements were not fully aware of this effect. Designers ignored this completely. By shifting towards lager "supports" (following the terminology of Habraken, ref.: 21) it became possible as well to offer (1) a variety of different units adapted to the preferences of individual end users, as to provide (2) a lot of opportunities for adaptation to future needs, as (3) to reduce costs.

7.4.4 The introduction of CAD/CAM

In the nineties this "Open Building System" approach proved to be a perfect preparation for the implementation of CAD/CAM. A case study (ref.: 20) describes in detail how based on a "3D CAD model" the following can be generated : all architectural drawings, working drawings used in the different stages of the production process, planning information, cost calculation, order lists for suppliers, files for steering NC-machines,....

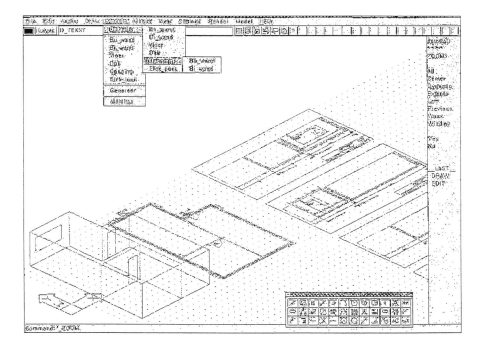

Figure 11

The model showed following characteristics:

- external and internal walls, floors and roofs were represented by rectangular planes, all of them on different layers (and thus with other colours);
- windows and doors are also represented as planes, but a lot of alpha-numerical data were associated, such as: type of frame, type of glazing, type of ironmongery, ...
- electrical ducts incorporated in the elements.

Macros enhanced the speed of constructing this 3D model (automatic switching to correct layers, systematic access to library of components, automatic association of alpha-numerical data with attributes or, in other cases, proposing defaults, ...). Possible inconsistencies were checked. Figure 11 shows (exceptionally on one drawing) the 3D model (left), generated architectural plans (middle) and the plans with the layout of the shutters on the production-tables (right).

7.4.5 Conclusion

As a conclusion one may say that the existence of a clearly defined "Open Building System" may free designers from an important amount of routine tasks, will increase the feedback of cost and other information from production and assembly to design, will allow an almost automatic steering of the production process, including the placement of just-in-time orders by suppliers.

7.5 SYSTEM THINKING FOR THE CONSTRUCTION SECTOR : A MORE GENERAL VIEW

The terms "Open" and "Closed Building Systems", as commonly used in the construction sector, are somewhat different from the more general definitions of a "system" and an "open system" in the field of "system thinking".

7.5.1 Open and closed systems in general

A "System" is generally defined as a set of entities interacting with each other and considered as a whole and distinct from the outside world. Each entity may be a system on its own, so a system may be composed of sub-systems.

E.g.: a heating system is composed of a heat generating system, a heat distributing system, a heat disseminating system and a control system.

An "Open System" is a system with links with the outside environment.

Those commonly accepted concepts are more general than the specific terminology widespread in the construction sector.

7.5.2 The system concept in the building sector

At least the following three concepts were used in the previous discussions within CIB W24 since its reactivation in May '95 (ref. :22-23):

1. Buildings (built or designed) as a system
2. Set of components (to be used in a building) as a system
3. Actors in the construction sector as a system.

The three concepts are described below:

7.5.2.1 Buildings (Built or Designed) as a System

For some people this is the most evident way of looking at a building. Depending on their background (facility manager, structural engineer, HVAC engineer ,...) different sub-systems are defined: structural system, circulation system, HVAC - system, thermal system, acoustical system, fire safety system, spatial system (spaces, sub-spaces, group of spaces, building, group of buildings,...),...

Most of the systems have common parts. For example a bearing wall will be part of the structural system, but is also subdividing space, is an acoustic barrier and is eventually delimiting a fire compartment.

7.5.2.2 Set of Components as a System

Complementary to the previous approach, where the final product "building" is considered (as built or just considering the final product in the design phase), it is evident that, when the building is being constructed, one is comparing different components which can be assembled. In many cases suppliers call this set of components a system :

curtain - wall component system, partition - wall component system, HVAC component system, sewerage component system, ...

By "system" one indicates that the different parts are studied in advance and designed in such a way that they can be combined in different ways.

Of course for each building different "sub-systems" have to be integrated.

7.5.2.3 Actors in the Construction Sector as a System

One of the main characteristics of the building sector is that in most projects several companies (architects, advisers, suppliers of products, contractors,...) have a specific input. The same combination of companies will only exceptionally collaborate in a next project.

Considering each actor separately, within his organisation several entities are interacting: workers on site, staff personnel, equipment,.... Co-ordinating all those inputs is time after time such a challenge that the management of construction projects became a specialised job.

For those people it is evident to see the actors (persons, organisations), who take part in the process, as the entities of a system.

A recent example of this approach looking at the present practice in many countries (Finland, France, Japan, The Netherlands) may be found in ref. 24.

7.5.3 Buildings as a system of "spaces" or a system of "material objects"

Along with the fact that the same entities (the wall in the example above) are part of so many different systems (structural, acoustic, fire control system,...), one of the major problems is that two completely different approaches are common:

- considering a building as a (hierarchical) system of spaces: rooms, cluster of rooms, building, cluster of buildings
  * e.g.: sleeping rooms, sleeping zone, apartment, apartment block, high-density residential cluster,...
  * patient rooms, nursing area, hospital wing, medical campus,...
- considering the material objects "delimiting" or "serving" those spaces, e.g.:
  * wall, roof, floor,...

* wall-panel, roof-tile, floor-beam, ...
* heating-system, liquid-supply-system, communication-system,...
* radiator, water-boiler, switch-box,...

As illustrated by the examples those components may be very specific: a specific radiator as part of a heating system, a specific wall-panel as part of a partitioning system, a three-dimensional bathroom component as part of the sanitary system,...

They may be considered "as built" or "as designed".

It is also possible to distinguish "conceptual systems" whereby only relevant aspects of certain components are considered.

* For a structural system this may be elasticity, specific mass, deformation,...
* For a thermal system important characteristics are: thermal transmission, thermal capacity, radiation,...

For those "conceptual systems" only the aspects considered to be relevant are taken into account. This creates the possibility to simulate behaviour without making final decisions about the selection of components.

For sure this has a lot of advantages. On the other hand, it is possible that conflicts and problems will arise later when concrete solutions are elaborated.

This "abstract" view of buildings is evident in design phase, but is also possible (by making abstraction of certain aspects) in an "as built" situation.

This approach of making abstraction of a lot of aspects of material objects is evident in system thinking. What is probably not so evident is the shift from physical objects to spaces, from material things to the "voids" defined by them.

However the description of buildings as a set of spaces is evident for the "end user". It is day-to-day practice, for the elaboration of a design brief. For the spaces a set of characteristics are defined: functional use, required temperature, humidity, sound level, illumination,...

When the building programme becomes more complex, a hierarchical set of spaces should be defined (ref. 25).

The requirements for the spaces have to be translated in performances for the "space delimiting" and the "space servicing" components. Products should be selected in function of those requirements.

In practice those approaches are often two worlds apart :

- end-users, people elaborating a programme brief, facility managers focusing on "spaces"
- contractors, people writing specifications, suppliers mainly considering "products" and the way they should be applied.

It is important to accept the different viewpoints and to link the two approaches. For both system-thinking can be useful.

7.5.4 Conclusions

System thinking has proved to be a powerful way of looking at reality, but one should avoid to use the same terms for different concepts and by thereby creating confusion. Therefore is seems to be good to accept the following three approaches:

7.5.4.1 Buildings as systems

The lifetime of a building is much longer that the process of design, production and construction. Therefore the simulation of the behaviour of buildings is foremost important. Buildings (only designed or built) should be seen as a system.

7.5.4.2 Set of building components as a system

Supply via a set of components, elaborated to fit together and with components of a third party, has proven to be efficient. Those sets of components, open for different combinations, are often called an "open (sub-)system". "Open Industrialisation" should create a framework for supply via "Open (sub-)systems".

7.5.4.3 Building sector as a system

The building sector may be seen as a system where actors, tools, products,... are linked to each other. Information exchange between different actors will become even more and more important. Therefore it should be possible to consider the construction sector as a system. The most important actor is the end user. Therefore adaptability to specific requirements and possible changes in the future (changeability, expandability, alternative use,...) are essential.

Recognising the different viewpoints when different people use the concept "building system" is a first step towards a better understanding. All of the approaches aim at a better output in relation to the input, at a better "productivity " of the total production chain, however difficult it may be to measure the (architectural) quality of the end product.

7.6 INFORMATION TECHNOLOGY : NEW OPPORTUNITIES FOR THE COLLABORATION BETWEEN DESIGNERS AND SUPPLIERS OF BUILDING SYSTEM

7.6.1 Problem description

The construction sector is composed of many "Small- and Medium-sized Enterprises". The scale of the enterprises is the first characteristic of the sector. A second characteristic has even more consequences : the fact that activities with very close links (like design and production) are spread over different firms.

In recent years most enterprises have implemented different software tools for drafting, for design, for bookkeeping, for cost estimation, for planning and scheduling, for administration,... Those programmes have been developed in most cases independently from each other. The data structure is elaborated in function of each application. The effort of updating and integrating those tools within each of the "Small- and Medium-sized Enterprises" (designers, producers, contractors, suppliers,...) absorbs so many resources that an approach integrating different enterprises with similar or complementary characteristics is hardly followed. Short-term-solutions within each enterprise are probably not in line with optimal long term developments for the whole sector.

The problem is very evident when it comes to the interaction between design and production of prefabricated concrete components : on one hand the design is elaborated without a clear insight in the possibilities and limitations of precasting units, on the other hand information technology is applied within the production sector on all levels without systematic feedback to design.

7.6.2 Basic transfer of electronic drawings

At this moment the most common exchange between designers, manufacturers and contractors is done via drawings and text <u>on paper</u> even if the drawings are generated via a CAAD-model and texts are manipulated with a word processor.

When exchanging "electronic drawings", there should be a correct <u>transfer</u> of drawing entities.

In most cases it is not sufficient that one "sees" the same result on paper before and after the transfer. This is certainly so if one analyses and edits the graphic result.

7.6.3 Need for updating drafting conventions

Even if the transfer of the graphical entities is correct the next question is if the same reality is represented by the transferred symbols, if there is a correct <u>interpretation</u> of drawing entities. This problem is in fact an old one, but as long as the interpretation was done by "human", "intelligent" actors, many ambiguities were solved based on an interpretation of symbols in their context.

There is not only a need for exchange of drawings at the end of the design process (when all details are fixed) but there is even more need for exchange from the first phases on when a lot of decisions are not yet taken. In a situation where "unfinished "models are exchanged, conventions are needed for representing parts of the design where only a few performances are fixed. Those conventions come on top of all other "scale-dependent" conventions.

Three concepts play a crucial role : modelling, phasing and integrating. The models should be adapted to the phase of the design process, they should be appropriate for each discipline and integration of different inputs should be possible.

7.6.4 Conventions for layering drawings : only a very first step

Recently a proposal for a new international standard (currently under discussion, ref.: 26) for "structuring layers in computer aided building design" has been published. The proposal may be very useful for improving the structure of CA(A)D drawings in the construction sector, but it is more looking for a combination of best practices in the past than for a future oriented approach.

The proposal is largely influenced by yesterday's and today's approach of drawing horizontal sections for each floor. A more future-oriented approach will raise the question which objects of building models will allow us to perform the desired checking on design proposals and that will allow us to exchange global or partial models.

7.6.5 Definition of modelling entities

The aim of each CA(A)D programme is to edit graphical entities and alpha-numerical entities. Between them all kind of links can exist. Most programmes developed for technical calculations have their own input module and store their data in a specific way. Changing the design means in that case inputting the changes in all the application programmes. Since the design process involves continuously changing (in an interactive graphical way) the characteristics of the provisional model of the building, it becomes clear that the search for a database shared between application programmes is a first priority for software developers.

Less ambitious than sharing the same database is the target of an automatic translation "to" and "from" a neutral format. If there are **n** application-programmes, **n(n-1)** transfer procedures have to be developed if one wants to exchange data directly from each programme to each other in the two directions. Via a neutral file format only **2n** transfer-procedures have to be elaborated.

STEP, the international **S**tandard for the **E**xchange of **P**roduct **D**ata, has been developed since the mid 80s (ref. : 27) . The initial release of the ISO standard was approved in March 1994. STEP uses a formal representation language (= EXPRESS) to specify the product information to be represented. STEP is very helpful in describing entities composed of basic or previously defined entities, with a lot of links, with inheritance, with derived properties, etc. ...But prior to that one has to decide what entities will be part of the model.

The fast growing applications of "object-oriented-programming techniques" have forced us to reflect on this problem in a very systematic way.

7.6.6 A conceptual model for buildings

The basic function of a building is to provide shelter, to offer enclosed spaces. Spaces may be grouped in clusters; those clusters in their turn in larger clusters. The level of nesting clusters depends on the size and type of building.

Spaces are enclosed by "space delimiting elements" like: external walls, internal walls, floors on the ground, internal floors, roofs,... Those elements are defined independently of the materials used in order to build them. A second group of elements is called "space servicing elements". They all play a role in providing the desired indoor environment: lighting, heating, ventilating, supplying water, evacuating sewerage, offering communication channels,... The two types of elements can be subdivided in smaller elements. Also those sub-elements are only defined by their function, by the (sub)-role they play in offering comfortable space to end users. Elements should be defined in such a way that a building is divided without "gaps and overlaps".

As well "buildings", "spaces" as "elements" are thus only defined by "function".

The requirements formulated for the spaces are translated to requirements for the elements.

Elements can only be constructed using one or more (building) products. All objects delivered on site in order to become parts of a building are called a "building product". A product may be an assembly of smaller products. The incorporation of similar products requires (on average) the same set of inputs: labour, equipment, secondary products, ... Those "applied products" will be called "Work sections". The concept of "Work sections" is used in bills of quantity, in lists with unit rates, as headings for specifications. A "Work section" may be composed of smaller "Work sections". "Work sections" and "products" are both physical objects. The required performances for the elements should be realised by a correct application of the appropriate building products.

"Products" and "Work sections" are characterised by a set of properties. All the physical objects may be subdivided in classes on different grouping principles: e.g.: constituent material (wood, concrete, steel,...), form (blocks, panels, pipes,...), properties (fire resistance, capillary, light reflection,...),... A data structure with multiple inheritance is evident.

The location of products is defined in relation to the location of a basic geometrical entity like : a point, a line, a surface or a volume. Those geometrical entities are a support for the designer in order to control the position and the size of the elements. The size and position of "Work sections" and "Products" is in the first place defined in relation to that basic geometrical entity. A further correction of the size and position may be needed based on the size or position of adjacent products. In modular co-ordination this problem is known as the "problem of the building node".

7.6.7 Conclusion

In the future every design will be elaborated with CAAD-tools. The enhanced exchange of information between the designer and units for the production of prefabricated components can improve the quality/cost-ratio. Exchanging parts of a 3D model (with geometrical and alpha-numerical information) should allow automatic deduction of characteristics of parametric components belonging to an open building (sub-)system. The exchange should be bi-directional and give the designer the opportunity to integrate the more detailed description of sub-parts or to rethink his model. The availability of new information channels can speed up the process significantly.

7.7 GENERAL CONCLUSIONS

The development of "Open Building Systems" within the context of "Open Industrialisation" creates the possibility for system suppliers to optimise their production and to control the integration of their products in the global process. Dimensional co-ordination remains a basic prerequisite. The emerging information technology can improve the exchange in the design phase between designers and suppliers of building systems. It becomes even more evident that a detailed description of "Building Nodes" is essential.

REFERENCES

1.   BLACH, K. AND MCEVATT, W.,
      The principles of modular co-ordination in building
      Dublin/Hørsholm, An Foras Forbartha/Danish Building Research
      Institute, 1984

2. ECONOMIC COMMISSION FOR EUROPE
    Dimensional co-ordination in building - Current trends and policies
    in ECE countries
    New York, United Nations, 1974
3. L'AGENCE EUROPÉENNE DE PRODUCTIVITÉ DE
    L'ORGANISATION EUROPÉENNE DE COOPÉRATION
    ÉCONOMIQUE
    La coordination modulaire dans le bâtiment
    Paris, AEP-OECE, 1956
4. L'AGENCE EUROPÉENNE DE PRODUCTIVITÉ DE
    L'ORGANISATION EUROPÉENNE DE COOPÉRATION
    ÉCONOMIQUE
    La coordination modulaire dans le bâtiment - Deuxième rapport
    Paris, AEP-OECE , 1991
5. IMG/CIB-W24
    Condensed principles of modular co-ordination
    in: Bulletin CIB, 1964, n° 3, p 14-16
6. Sine autor
    1983 Compendium of CIB working commissions and steering
    groups
    Rotterdam, CIB Report - Publication 71, 1983
7. SCHMID, THOMAS, TESTA, CARLO
    Systems building- Bauen mit Systemen - Constructions modulaires
    Zürich, Artemis, 1969
8. DELRUE,J
    Architecturale grondslagen voor een rationalisatie in de
    bouwnijverheid
    Modulaire maatcoördinatie als ontwerpmethodiek
    (Architectural basis a rationalisation in the building sector
    Modular co-ordination as design methodology)
    Leuven, Ph. D. Dissertation Univ. Leuven, 1969
9. DIETZ, ALBERT G. H.
    Industrialized building systems for housing
    Cambridge, MIT Press, 1971
10. NISSEN, HENRIK
    Industrialized building and modular design
    London, Cement and concrete association, 1972
11. LUGEZ, JEAN, BACHÈRE, G.
    La préfabrication lourde en panneaux et le bâtiment d'habitation
    Paris, Eyrolles, 1973

12. BLACHÈRE, GÉRARD
    Technologies de la construction industrialisée
    Paris, Eyrolles, 1975
13. ACC (Association pour la Construction en Composants)
    Conventions de coordination dimensionnelle: Conventions Générales
    Paris, Edition du Moniteur, 1978
14. BERNARD, PAUL
    La construction par composants compatibles
    Paris, Moniteur, 1980
15. RUSSELL, BARRY
    Building systems, industrialization, and architecture
    London, Wiley, 1981
16. ISO
    Modular co-ordination - Vocabulary-Coordination modulaire -
    Vocabulaire
    ISO, international standard 1791 - 1973
17. IMG/CIB W24
    Condensed principles of modular co-ordination
    in: Bulletin CIB, 1964, nr. 3, p. 14 - 16
18. DELRUE, J. and DE TROYER, F.,
    Open system building with concrete panels for housing-Design guide
    for architects
    in: Proceedings of the CIB symposium on "System Building"
          Budapest, Institute for Building Science ETI, 1981, p. 427-442.
19. DE TROYER, F.,
    How to overcome the information gap between the producers of
    concrete elements and the designers of housing.
    in: Proceedings of the third international Symposium on
    Organisation and
          Dublin, An Foras Forbartha, 1981, p. A.2.117-129.
20. DE TROYER, FRANK,
    Architectural quality and lean production
    Reading, CIB W24, 6 Nov 1995
21. HABRAKEN, N.J.,
    Supports an alternative to mass housing
    1972, The Architectural Press, London
22. DE TROYER, FRANK,
    OPEN INDUSTRIALISATION IN BUILDING :
    Definition of objectives, concepts an the role of dimensional co-
    ordination
    Helsinki, CIB W24, 22 May 1995

23. DE TROYER, FRANK,
    Concept and terminology concerning "Open Industrialisation"
    Leuven, CIB W24, June 3-4, 1996
24. LAHDENPRÄ, P.,
    Reorganizing the building process - the holistic approach
    Espoo, VTT publications, 1995
25. SARJA, A. & HANNUS, M.,
    Modular systematics for the industrialized building
    Espoo, VTT publications, 1996
26. ISO DIS 13567 - The proposed international standard for structuring
    layers in computer aided building design
    http://itcon.org/1997/2/paper.htm
27. ISO 10303 STEP (Standard for the Exchange of Product Data)
    http://www.iso.ch/liste/TC184SC4.html

Keyword Index

This index has been compiled from the keywords assigned to the chapters, and edited and extended as appropriate. The numbers refer to the first page of the relevant chapters.